Geology of the (
Worcester

The district described in this memoir is one of great geological interest, encompassing parts of the Pre-cambrian inlier of the Malvern Hills, the Lower Palaeozoic Welsh Borderland and the Permo-Triassic Worcester Basin. The memoir brings together, in a systematic way, the results of over 150 years of geological research in one of the classical areas of British geology. It provides an up-to-date assessment of the geology, synthesising the results of recent BGS studies which have led to increased understanding of the nature of the Precambrian basement and of the evolution of the Worcester Basin.

The memoir chronicles the district's geological history, from Proterozoic accretion on to the northern margins of the southern hemisphere continent of Gondwana, the rifting and northward drift of the East Avalonian plate through the Cambrian, Ordovician and Silurian, its collision and docking with the North American Laurentian and European (Baltica) plates during the Acadian Orogeny, tectonic inversion of the region during the Variscan Orogeny, and finally, Permo-Triassic rifting and the formation of the Worcester Basin. Superimposed on these tectonic events, global sea-level rises are documented by the transgressive marine deposits of Cambrian, late Llandovery, early Ludlow and Rhaetian age. The discovery of two new interglacial deposits of pre-Ipswichian age adds to the emerging picture of climatic changes that affected Britain in the Middle Pleistocene.

Current economic interest is mainly centred on the district's groundwater and sand and gravel resources. Over 21 million litres of Malvern Water are bottled annually and the terrace deposits are a large potential aggregate resource close to the main urban centres of the west Midlands.

The research which this memoir summarises, and the new geological maps which it describes, are a valuable addition to the great volume of geological literature that the area has generated, and provide a sound basis for future research, land-use planning, development and conservation of one of the most beautiful parts of the country.

Cover photograph
The north Malvern Hills, looking north-north-east from Broad Down. The Severn valley, underlain by the Worcester Basin, lies to the east.

Herefordshire Beacon viewed from the north.

BRITISH GEOLOGICAL SURVEY

W J BARCLAY
K AMBROSE
R A CHADWICK
T C PHARAOH

CONTRIBUTORS

Stratigraphy and structure
A J M Barron
A Brandon
A A Jackson
B S P Moorlock
P A Rathbone
R J Wyatt

Biostratigraphy
R J Aldridge
H F Barron
H C Ivimey-Cook
J E Mabillard
G Warrington
D E White

Mineral resources
P M Harris

Geochemistry
T S Brewer

Geophysics
J D Cornwell

Water resources
C S Cheney

Petrology
R A Dearnley

Geology of the country around Worcester

Memoir for 1:50 000 Geological Sheet 199
(England and Wales)

London: Stationery Office   1997

First published 1997

ISBN 0 11 884512 8

Bibliographical reference

BARCLAY, W J, AMBROSE, K, CHADWICK, R A, and PHARAOH T C. 1997. Geology of the country around Worcester. *Memoir of the British Geological Survey*, Sheet 199 (England and Wales).

Authors

W J Barclay, BSc
K Ambrose, BSc
R A Chadwick, MA, MSc
T C Pharaoh, BSc, PhD
British Geological Survey, Keyworth

Contributors

A J M Barron, BSc
H F Barron, BSc, MSc
A Brandon, BSc, PhD
J D Cornwell, MSc, PhD
A A Jackson, BA, PhD
B S P Moorlock, BSc, PhD
G Warrington, BSc, PhD
British Geological Survey, Keyworth

C S Cheney, MSc
British Geological Survey, Wallingford

R A Dearnley, BSc, PhD
P M Harris, MA
H C Ivimey-Cook, BSc, PhD
R A Old, BSc, PhD
P A Rathbone, BSc, PhD
D E White, MSc, PhD
R J Wyatt, BSc
formerly British Geological Survey

R J Aldridge, BSc, PhD
The University, Leicester

T S Brewer, BSc, PhD
The University of Nottingham

J E Mabillard, BSc, PhD
formerly The University of Nottingham

Printed in the UK for the Stationery Office
Dd 296605   C6   2/97

Other publications of the Survey dealing with this and adjoining areas

BOOKS

Memoirs
Geology of the country around Droitwich, Abberley and Kidderminster, Sheet 182, 1962
Geology of the country around Redditch, Sheet 183, 1991
Geology of the country between Leominster and Hereford, Sheet 198, 1989
Geology of the country around Stratford-upon-Avon, Sheet 200, 1974
Geology of the country around Tewkesbury, Sheet 216, 1989

British Regional Geology
The Welsh Borderland, 3rd edition, 1971
Central England, 3rd edition, 1969

MAPS

1:1 000 000
Pre-Permian geology of the United Kingdom (South), 1985
Geology of the United Kingdom, Ireland and adjacent continental shelf, south sheet, 1991

1:625 000
Geological map of the United Kingdom, south sheet, 3rd edition, 1979
Quaternary map of the United Kingdom, South, 1977
Bouguer anomaly map of the British Isles, south sheet, 1986
Aeromagnetic map of Great Britain, south sheet, 1965
Hydrogeological map of England and Wales, 1977

1:250 000
Solid geology, mid-Wales and Marches, 1990
Bouguer gravity anomaly, mid-Wales and Marches, 1986
Aeromagnetic anomaly, mid-Wales and Marches, 1980

1:50 000
182 (Droitwich) Solid and Drift (1976)
183 (Redditch) Solid and Drift (1989)
198 (Hereford) Solid and Drift (1989)
200 (Stratford-upon-Avon) Solid and Drift (1974)
216 (Tewkesbury) Solid and Drift (1988)
217 (Moreton-in-Marsh) Drift (1981)

CONTENTS

FIGURES

PLATES

TABLES

PREFACE

This memoir describes the geology of the district covered by the 1:50 000 Scale New Series Sheet 199 (Worcester) of the Geological Map of England and Wales. The district is one of great geological interest, encompassing parts of the Precambrian inlier of the Malvern Hills, the Lower Palaeozoic Welsh Borderland and the Permo-Triassic Worcester Basin. The memoir brings together, in a systematic way, the results of over 150 years of geological research in one of the classical areas of British geology. It provides an up-to-date assessment of the geology, synthesising the results of the recent BGS studies which have led to increased understanding of the nature of the Precambrian basement and of the evolution of the Worcester Basin.

Our studies have included detailed geological mapping, geochemical basement studies, seismic reflection traverses, aeromagnetic and gravity analyses, geothermal assessment and the drilling of the Department of Energy's deep borehole at Kempsey.

The detailed lithostratigraphy and chronostratigraphy chronicle the district's geological history, from Proterozoic accretion on to the northern margins of the southern hemisphere continent of Gondwana, the rifting and northward drift of the East Avalonian plate through the Cambrian, Ordovician and Silurian, its collision and docking with the North American Laurentian and European (Baltica) plates during the Acadian Orogeny, tectonic inversion of the region during the Variscan Orogeny, and finally, Permo-Triassic rifting and the formation of the Worcester Basin. Superimposed on these tectonic events, global sea-level rises are documented by the transgressive marine deposits of Cambrian, late Llandovery, early Ludlow and Rhaetian age.

Our knowledge of the Quaternary history of the region has been greatly increased by the discovery of two new interglacial deposits of pre-Ipswichian age, adding to the emerging picture of climatic changes that affected Britain in the Middle Pleistocene.

Current economic interest is mainly centred on the district's groundwater and sand and gravel resources. Schweppes bottles over 21 million litres annually of Malvern Water from a spring at Colwall, worth about £11 million at 1990 prices. The mapping of over 200 km^2 of the district was funded by the Department of the Environment as part of its land-use planning programme and the need to delineate resources for the construction industry within easy reach of the principal conurbations of the Midlands. The Kempsey Borehole disproved the previously held view that Coal Measures might be present beneath Worcester, but our seismic studies have demonstrated the presence of concealed Carboniferous rocks to the east. These studies have also suggested the presence of buried salt deposits in southward continuation of the Droitwich deposits. The combination of the discovery of Precambrian basement to the Worcester Basin precludes oil prospectivity in the basin. Likewise, in the Lower Palaeozoic terrain, the Collington Borehole and our geophysical studies indicate low oil prospectivity over much of the district, Cambrian and Ordovician source rocks being absent.

I believe that our recent work in and around the Worcester district has added greatly to geological knowledge, and that this memoir is a valuable addition to the great volume of geological literature that the area has generated. I am also confident that the work provides a sound basis for future research, and for land-use planning, development and conservation of one of the most beautiful parts of Britain.

Peter J Cook, CBE, DSc, CGeol, FGS
Director

British Geological Survey
Kingsley Dunham Centre
Keyworth
Nottingham NG12 5GG

ACKNOWLEDGEMENTS

This memoir has been compiled by Mr W J Barclay. The authorship of each chapter is as follows:

One	Introduction	W J Barclay
Two	Precambrian	T C Pharaoh and W J Barclay
Three	Cambrian and Ordovician	W J Barclay
Four	Silurian	W J Barclay and D E White*
Five	Devonian	W J Barclay
Six	Carboniferous	W J Barclay
Seven	Permian and Triassic	K Ambrose, H C Ivimey-Cook* and G Warrington*
Eight	Jurassic	K Ambrose, H C Ivimey-Cook* and G Warrington*
Nine	Quaternary	W J Barclay and B S P Moorlock†
Ten	Structure	R A Chadwick, W J Barclay, and J D Cornwell‡
Eleven	Economic Geology	C S Cheney§, P M Harris‖ and W J Barclay

* Biostratigraphy
† Terrace Deposits
‡ Geophysical potential-field investigations
§ Water resources
‖ Mineral resources

In addition to the authors listed above, Two (Precambrian) incorporates petrographical analyses carried out by Dr R A Dearnley, and Mr H F Barron provided palynological identifications of samples from the Silurian. Open-file BGS Technical Reports (Appendix 1) by Mr A J M Barron, Drs A Brandon, A A Jackson, R A Old and P A Rathbone and Mr R J Wyatt have been used in writing this account. The memoir was edited by Mr T J Charsley and Dr A A Jackson.

Grateful acknowledgement is made to organisations and individuals, including landowners, quarry operators and public and local authorities, for their willing help and cooperation during the course of the geological and geophysical surveys. We are indebted to Dr T S Brewer for contributing his chemical analyses of the rocks of the Malverns Complex. Chemical analysis of the rocks collected during the survey was carried out by the University of Nottingham. Dr R J Aldridge of Leicester University and Dr J E Mabillard, formerly of the University of Nottingham, kindly provided a contribution on Silurian microfaunal assemblages and Dr J S W Penn of Kingston University supplied Plate 3.

The geological survey of sheets SO 84 and 85 and parts of sheets SO 64 and SO 65 was partly funded by the Department of the Environment.

NOTES

Throughout this memoir the word 'district' refers to the area included in the 1:50 000 geological sheet 199 (Worcester).

Figures in square brackets are National Grid references. Those not preceded by two letters lie within the 100 km grid square SO.

Numbers preceded by the letter E refer to the BGS sliced rock collection; those preceded by A refer to the Survey's photograph collection.

The authorship of fossil names is given on pp.146–148.

Enquiries concerning geological data for the district should be addressed to the Manager, National Geological Records Centre, BGS, Keyworth. Geological advice on the area should be sought from the Regional Geologist, Central England and Wales, BGS, Keyworth.

This memoir provides a general account of the geology of the district, with selected details only. For fuller details of sections and exposures the reader is referred to the BGS Technical Reports listed in Appendix 1.

All localities mentioned in this memoir are either privately owned or are administered by the Malvern Hills Conservators. Permission for access should be obtained from landowners. Hammering is not permitted in the Malvern Hills and permission for collecting samples must be obtained from the Conservators. The Code of Conduct published by the Geologists' Association should be adhered to.

HISTORY OF SURVEY OF THE WORCESTER SHEET

The district covered by the Worcester (199) sheet of the 1:50 000 geological map of England and Wales was originally surveyed at the one-inch scale by W T Aveline, H H Howell, E Hull, J Phillips and A R Selwyn as part of Old Series sheets 43, 44, 54 and 55 and published between 1845 and 1856. A memoir by Phillips describing the geology of the Malverns–Abberley area was published in 1848.

Parts of the east of the district were surveyed at the six-inch scale by W A E Ussher in 1885–86 on County Series sheets Worcs. 29SW, 29SE, 40NE, 40SE, and 41NW, NE, SW and SE. In the course of examining the new Colwall railway tunnel in 1924–25, T Robertson surveyed part of the north Malverns, on County Series sheets Worcestershire 39SE (south-east part) and 46NE (eastern part). BGS also holds the field maps of the Malverns by A H Green, prepared subsequent to his resignation from the Survey in 1874.

The northern margin of the district, north of grid line 58, was surveyed on the six-inch scale by S E Hollingworth, J H Taylor and T N George, as overlap from the Droitwich (182) sheet on County Series sheets Herefordshire 14SW and 14SE, and Worcestershire 27SW, 27SE, 28SW, 28SE, 29SW and 29SE between 1931 and 1936. The eastern edge of the district was surveyed at the same scale on National Grid sheets by A Whittaker and B J Williams in 1966–67 as overlap from the Stratford (200) sheet. Surveying of the main part of the Worcester sheet at the 1:10 000 scale was carried out between 1982 and 1987, following surveys of the Hereford (198) sheet to the west and Tewkesbury (216) sheet to the south, both of which included small overlap areas onto the Worcester sheet.

Geological 1:10 000 scale National Grid maps included wholly or partly in the 1:50 000 Worcester sheet are listed below, together with the initials of the geological surveyors and dates of survey; in the case of marginal sheets, all surveyors are listed. The surveyors were: K Ambrose, W J Barclay, A J M Barron, A Brandon, A A Jackson, B S P Moorlock, R A Old, P A Rathbone, P J Strange, A Whittaker, B J Williams, B C Worssam and R J Wyatt.

Copies of the fair-drawn or manuscript maps may be inspected at BGS Keyworth and purchased as black and white dyeline sheets.

SO 63 NE	Little Marcle	BSPM	1981
SO 64 NE	Bishop's Frome	WJB, PJS	1981, 1987
SO 64 SE	Bosbury	AB	1980–81
SO 65 NE	Tedstone Delamere	WJB, AAJ	1983
SO 65 SE	Stanford Bishop	PJS, PAR	1981, 1986–87
SO 73 NW	Ledbury	BSPM	1979–81
SO 73 NE	South Malvern Hills	BSPM	1979
SO 74 NW	Cradley	WJB	1985, 1987
SO 74 NE	Great Malvern	WJB	1982, 1987
SO 74 SW	Coddington	AB	1985–86
SO 74 SE	Malvern Wells	BSPM	1982
SO 75 NW	Whitbourne	WJB, PAR	1984–85
SO 75 NE	Kenswick	PAR	1984–85
SO 75 SW	Suckley	PAR	1985–86
SO 75 SE	Leigh Sinton	AJMB, PAR	1983, 1985
SO 83 NW	Longdon	BSPM	1979
SO 83 NE	Ripple	BSPM, RJW	1980
SO 84 NW	Madresfield	BSPM	1983
SO 84 NE	Kempsey	KA	1983
SO 84 SW	Hanley Swan	BSPM	1983
SO 84 SE	Earls Croome	KA	1983
SO 85 NW	Worcester NW	KA, AJMB, AB	1983–84
SO 85 NE	Worcester NE	BSPM	1984
SO 85 SW	Worcester SW	AJMB	1983
SO 85 SE	Worcester SE	AJMB	1983
SO 93 NW	Bredon and Ashchurch	AW, RJW	1967, 1979–80
SO 93 NE	Beckford	AW, BCW	1966, 1981
SO 94 NW	Pershore	RAO	1985–86
SO 94 NE	Wyre Piddle	KA, AW, RAO	1966, 1985–86
SO 94 SW	Eckington	RAO, AW	1966, 1985
SO 94 SE	Comberton	RAO, AW	1966, 1985–86
SO 95 NW	Crowle	KA	1985
SO 95 NE	Grafton Flyford	KA, BJW	1966, 1983, 1985
SO 95 SW	White Ladies Aston	KA	1985
SO 95 SE	Bishampton	BJW, KA	1966, 1985

ONE

Introduction

This memoir describes the geology of the area covered by the Worcester 1:50 000 Geological Sheet (199), published as a solid with drift edition in 1993.

The Worcester district encompasses some of the most delightful, unspoilt, rural countryside in England. It falls into two distinct areas separated by a north–south ridge of Lower Palaeozoic rocks which culminates in the inlier of Precambrian rocks that forms the Malvern Hills in the south of the district (Figures 1 and 2). To the west lie the undulating farmlands of Herefordshire, underlain mainly by the red beds of the late Silurian to Devonian Old Red Sandstone. The Malvern Hills comprise a narrow ridge that rises to a height of 425 m, steeply and dramatically from the Severn plain in the east, less so from the ground to the west. The northern 8 km of the ridge lie within the district, a further 5 km lying to the south. The area of the Malverns is a designated Area of Outstanding Natural Beauty. Consisting mainly of Precambrian intrusive igneous rocks, with some volcanic and volcaniclastic rocks, the hills lie on a major north–south structural lineament, the Malvern Axis, which is one of the most important basement structures of southern Britain. To the north of the hills, the structure is marked by an anticline with Llandovery rocks exposed along its core. To the east of the axis lies the Worcester Basin, a Permo-Triassic graben filled with up to 3 km of red-bed sediments, and bounded in the west by the East Malvern Fault. These rocks underlie the Severn valley and the subdued agricultural area in its vicinity. To the east, basal beds of the Lias form an escarpment and gentle dip slopes, with the bold escarpment of Bredon Hill lying in the extreme south-east.

The River Severn provides the main drainage of the district, the River Teme being its principal tributary (Figure 2). A short stretch of the River Avon is present in the south-east.

The cathedral city of Worcester, with a population of 75 000, is the largest settlement. Situated at an early crossing point on the River Severn, the site was settled first in the Bronze Age and later occupied by the Romans. The foundations of the present city were laid during the late Saxon and Norman periods. Industry and commerce developed in mediaeval times when the city was a river port. Its main expansion took place during the Industrial Revolution in the nineteenth century, mainly due to the construction of the Worcester–Birmingham canal, which provided a link with the major industrial development in the west Midlands. The city's traditional industries — glove, porcelain and sauce manufacture — have more recently been supplemented by a

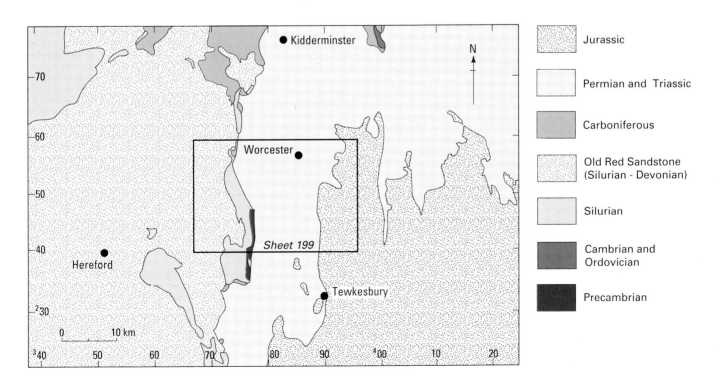

Figure 1   Geological setting of the district.

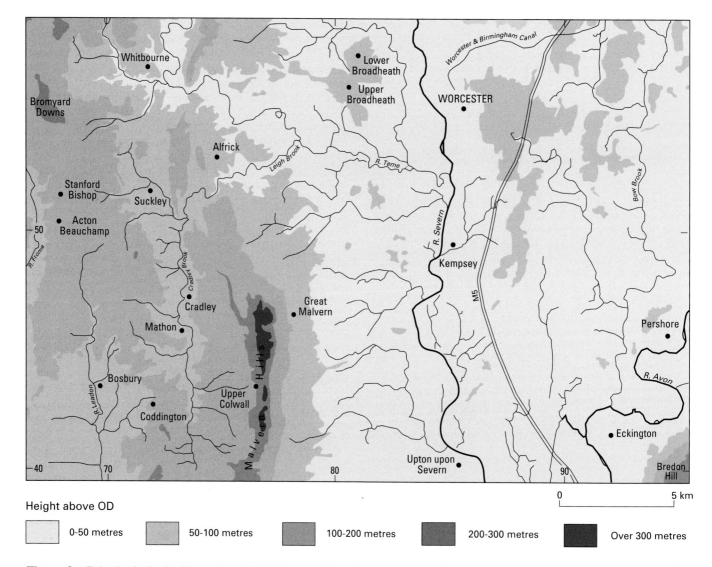

Height above OD

0-50 metres	50-100 metres	100-200 metres	200-300 metres	Over 300 metres

Figure 2 Principal physical features and drainage of the district.

range of light, mainly engineering industries, largely due to the proximity of the west Midlands' car manufacturing and components industry. These provide over sixty per cent of the employment of the workforce. The city is today the administrative centre of the county of Hereford and Worcester, and a regional service centre. For a fuller account of the city and its origins, the reader is referred to Barker (1969), McKee (1973) and Minshull (1974).

Great Malvern has a population of 30 000 and is the largest of a group of towns and villages strung around the flanks of the Malvern Hills. With a commanding view of the Severn valley, it is a popular tourist centre which expanded rapidly as a Victorian spa town, when visitors came to take the pure waters emanating from springs in the hills. Its former inhabitants include Edward Elgar who drew inspiration for his music from the beauty of the area. Ministry of Defence research laboratories played a prominent role in the development of radar during the second world war and are today Great

Malvern's largest employer. The Morgan car factory is one of several light industrial establishments. Private schools and retirement homes are a feature of the town. Pershore and Upton-on-Severn are among the numerous small towns and villages scattered throughout the district. The village of Colwall produces the district's most valuable commodity, Malvern Water, over 21 million litres of which are bottled annually by Schweppes.

The Malvern Hills are one of the classical areas of English geology. They have long attracted geological interest, since the work of Horner (1811), Murchison's visit in the 1830s and the first geological survey by Phillips in 1842. A large geological bibliography has accumulated, including three guide books (Hardie, 1969; Penn and French, 1971; Bullard 1989). Controversy and debate has centred on the origins of the Malverns, their structure, and relationships to the surrounding rocks. The area, with its geological diversity, wide age range of rocks and good level of exposure continues to attract large numbers of geologists and students.

GEOLOGICAL SUCCESSION

The succession of rocks and superficial deposits present in the district is shown inside the front cover. The geological setting of the district is shown in Figure 1.

GEOLOGICAL HISTORY

The Precambrian igneous and volcanic rocks of the Malvern Hills are about 681 Ma (e.g. Tucker and Pharaoh, 1991) and are believed to have formed during south-easterly directed subduction of oceanic lithosphere and accretion on to the northern margins of the southern hemisphere continent of Gondwana (Thorpe et al., 1984). The Iapetus Ocean lay to the north. The Malverns Complex is a calc-alkaline magmatic suite that underwent at least one high-grade, regional, metamorphic event, accompanied by ductile shearing, in the Proterozoic. Later brittle deformation is probably of Variscan age. The Warren House Formation is about 566 Ma (Tucker and Pharaoh, 1991) and is probably a tectonic sliver emplaced during Proterozoic movement of the East Malvern Fault. The volcaniclastic rocks of the Kempsey Formation, which form the basement to the Worcester Basin within the district, are similar geochemically to those of Charnwood, suggesting that the Charnwood Terrane extends as far west as the East Malvern Fault, which was thus a Proterozoic suture (Pharaoh et al., 1991).

Little can be deduced of Cambrian history from the small amount of Cambrian rocks within the district, although rifting of the Eastern Avalonia plate from Gondwana had probably begun by the end of the Precambrian (e.g. Soper and Woodcock, 1990). The Malvern Quartzite rests unconformably on the Malverns Complex to the south and contains clasts of Malverns Complex lithologies and volcanic (possibly Warren House Formation) rocks, suggesting that sediment was being dispersed from an adjacent terrane and deposited in a shallow marginal marine setting (Worssam et al., 1989). The White-leaved Oak Shale represents deeper-water, euxinic deposition in the developing Welsh Basin. As the Iapetus Ocean closed through the Ordovician and Silurian, East Avalonia tracked northwards to collide with the Laurentian and Baltican continents. The detailed timing and nature of the collision remain a matter of debate. However, inversion during the Ordovician resulted in uplift and exposure of the district. The predominant north-west–south-east structural grain of the Silurian and Devonian rocks of the district is thought to be a result of later reactivation of the structural template formed at that time.

Sedimentation recommenced in the late Llandovery, probably as a result of glacioeustatic global sea-level rise, when the Malverns formed the eastern shoreline of the Welsh Basin and the western margin of the Midland Platform. The district remained marginal to the basin throughout the Silurian, during which several transgressive–regressive cycles resulted in deposition of an intercalated carbonate–siliciclastic, shallow-marine, shelf sequence. Final closure of Iapetus in the late Silurian resulted in deposition of the Old Red Sandstone continental red-bed sequence, the molasse deposits of the late Caledonian (Acadian) Orogeny. Final docking of East Avalonia with Laurentia took place in the Middle Devonian when the region was inverted and underwent erosion.

There is little stratigraphical evidence from within the district of the events from the early Devonian until late Carboniferous, with Variscan inversion leading to the stripping of the Palaeozoic cover sequence. The Malvern Axis was a major controlling influence at that time, with westerly directed thrusting and dextral transpressive movement resulting in the present configuration of the Lower Palaeozoic rocks west of the East Malvern Fault. Small amounts of Westphalian coal measures are preserved on the axis, but there are no Carboniferous rocks preserved beneath the Worcester Basin within the district.

Extension during the Permian and Triassic reactivated the westerly directed Variscan thrusting, the East Malvern Fault forming the western margin of the Worcester Basin. The basin is a roughly symmetrical graben in which up to 3 km of red beds accumulated. Late Triassic sea-level rise resulted in deposition of shallow-marine Rhaetian and early Jurassic sequences.

Uplift and erosion of the district in the Cenozoic was caused by Alpine collisions to the south. Subsequent regional uplift and eastward tilting continues to the present day. There is, however, no stratigraphical record of the events that affected the district from the early Jurassic until the Middle Pleistocene.

The glacigenic deposits of the district are attributed to the Anglian, with the Mathon Sand and Gravel perhaps being the deposits of a pre-glacial river. An interglacial (pre-Anglian) silt is present beneath these gravels. Later Pleistocene deposits comprise temperate (possibly Hoxnian) silts and cold-stage solifluction and gelifluction deposits, as well as a suite of well-developed river terrace deposits.

TWO

Precambrian

Precambrian rocks crop out in the Malvern Hills, occupying seven fault-bounded blocks in a narrow, north–south–trending ridge about 13 km long (Figure 3). The northernmost 8 km lie within the Worcester district. The strongly banded nature of many of the Precambrian outcrops led many of the early researchers (e.g. Horner, 1811; Holl, 1865) to propose an origin by metamorphism of sedimentary rocks. The volcanic origin of the rocks to south-east of Herefordshire Beacon (Warren House Formation) was recognised by Holl (1865) and confirmed by Callaway (1880) and Green (1895). Callaway (1889) recognised that the remainder of the Precambrian metamorphic rocks (Malverns Complex) were derived from a plutonic igneous complex by strong ductile and brittle shearing. Petrographical and geochemical investigations (e.g. Lambert and Holland, 1971; Thorpe, 1972) have confirmed this interpretation. Metasedimentary rocks, if any, make up less than 1 per cent of the exposed area of the complex.

Precambrian volcaniclastic rocks, here named the Kempsey Formation, have been proved in the basement of the Worcester Basin. Table 1 summarises the geochronology of Precambrian events of the district.

MALVERNS COMPLEX

The name Malverns Complex (Lambert and Holland, 1971) is preferred in this account to Malvernian Complex (Worssam et al., 1989), in order to avoid the chronostrati-

Table 1 Chronology of Precambrian events in the Worcester district.

Age	Terrane		
	MALVERNS COMPLEX	WARREN HOUSE FORMATION	BASEMENT OF WORCESTER BASIN
Latest Precambrian <560 Ma	Amalgamation of Avalonian accretionary terrane of southern Britain		
	Greenschist facies; prograde in dykes, retrograde in Malverns Complex	Greenschist facies	?Greenschist facies
c.565 Ma	Emplacement of microdiorite dykes		
566 ± 2 Ma[1]		Basaltic–andesite lavas and felsic tuffs erupted in oceanic marginal basin	
>603 Ma			Deposition of volcaniclastic sediments of the Kempsey Formation, fed from Charnian arc
Late Precambrian			
c.600 Ma[2]	Uplift of Malverns Complex; ?Pseudotachylites Pegmatites Trachyte body		
614–610 Ma[2]	Mylonitic fabrics, ductile shearing		
650 Ma[4]	Metamorphism in almandine-amphibolite facies		
670 ± 10 Ma[3]	Granites, tonalites		
677 ± 1 Ma[1]	Plutonic complex		
?>677 Ma	Gabbros, diorites		

1 U-Pb zircon age (Tucker and Pharaoh, 1991)
2 K-Ar minimum muscovite or biotite age (Fitch et al., 1969)
3 U-Pb monazite age (Thorpe et al., 1984)
4 Ar-Ar hornblende isotope correlation ages (Strachen et al., 1996).

graphical implications of the latter. Geochemical analyses of the rocks are presented as Appendix 2.

The Malverns Complex crops out in the Malvern Hills, including North Hill [770 465], Sugarloaf Hill [765 458], Worcestershire Beacon [769 452], the long ridge west of Malvern Wells, Black Hill [766 405] and Herefordshire Beacon [760 400]. There are also two small inliers [7617 4795; 7625 4766] at Cowleigh Park.

The complex is composed largely of diorite, tonalite and hybridised and sheared derivatives of these protoliths (Figure 3). In most exposures a penetrative foliation is present, locally giving the plutonic protolith a schistose or superficially 'gneissic' appearance. Phyllonitic and mylonitic rocks are also present, but the intensity of deformation in the northern Malvern Hills is, in general, less than farther south (Worssam et al., 1989). Mafic and ultramafic rocks are present in minor amounts. Granite occupies the north-western and southern slopes of Worcestershire Beacon, exhibiting a complex relationship to the diorite host (Bullard, 1975; 1989), and is also present in smaller intrusive bodies and sheets elsewhere. Discordant pegmatites postdate most of the ductile deformation, as do mafic bodies on the eastern slopes of North Hill and Worcestershire Beacon. Thin sheets of microdiorite are also discordant to the ductile foliation of the complex.

The most detailed petrography of the complex is that of Lambert and Holland (1971). Diorites and tonalites form about 75 per cent of the outcrop, typically comprising green hornblende and tabular plagioclase (andesine). Relict igneous textures showing interlocking and ophitic feldspar laths and plates, zoned plagioclase feldspars and pseudomorphs after olivine are found in a few specimens of the more mafic diorites (Dearnley, 1990). Accessory minerals include sphene, apatite and zircon. With increasing grade of deformation and metamorphism, hornblende is replaced by epidote and chlorite, and oligoclase is progressively sericitised, ultimately producing the rock type referred to as 'epidiorite' by Blyth and Lambert (1970) and Lambert and Holland (1971). The range of dioritic rocks is most easily demonstrated in the North Hill Quarries [769 470; 771 469; 772 468]. Ultramafic rocks, consisting largely of hornblende with subsidiary biotite, are found in small amounts in many of the quarries, most notably at Tollgate Quarry [770 441], where a hornblende-pyroxenite mass is invaded by coarse granite pegmatites. Some of the ultramafic rocks may have a cumulate relationship to the more chemically evolved rocks, others forming from hybridisation reactions (Lambert and Holland, 1971).

Massive, well-jointed, pink granite is exposed in Earnslaw [771 445] and Tollgate quarries, although minor bodies and sheets are very widespread. The granites typically contain tabular microcline, oligoclase and biotite (<10 per cent); small amounts of muscovite, iron-

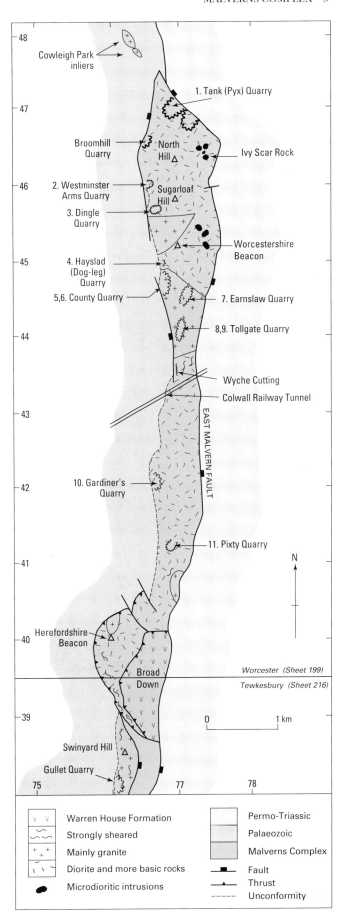

Figure 3 Simplified geological map of the Malverns Complex of the north Malverns. Lithologies of the complex are after Lambert and Holland (1971). Numbered localities are listed on p.6.

rich epidote, chlorite and iron-oxides make up the remainder, particularly in the more deformed examples, which also exhibit quartz aggregates and small irregular feldspar aggregates. The granite is variably foliated, ranging from virtually undeformed through to quartzofeldspathic mylonites (e.g. E58255 from Gullet Quarry [7625 3808] and E58268 from Hollybush Quarry [7593 3718]). A well-developed lineation is commonly present on some foliation surfaces. Bullard (1975), following Brammall and Dowie (1936), has advocated the presence of two distinct phases of granite emplacement.

Pegmatites are abundant throughout the complex, ranging from diffuse streaks in diorite to sharply discordant, dyke-like bodies. Potassic (microcline-quartz), sodic (albite-quartz) and granitic varieties have been recognised (Lambert and Holland, 1971). Muscovite is the principal accessory, together with minor amounts of biotite, secondary chlorite and epidote. Emplacement of granitic and pegmatitic bodies into the complex resulted in modification of the latter by hybridisation and metasomatism. Extensive hybridisation between diorites and felsic melts has been documented by Blyth and Lambert (1970). Metasomatism, involving the elements Na, Ca, K, Mg, Fe and Ti, is most clearly demonstrable in the ultramafic protoliths, such as the biotite amphibolites.

A small potash-rich intrusive body of trachytic composition in Earnslaw Quarry [771 445] (Thorpe, 1971) is affected by the same ductile deformation as the rest of the complex. Thorpe suggested that the unusual magmatic liquids from which it crystallised might have resulted from partial melting of biotite-rich components of the complex.

Principal quarry exposures

The following lists the principal quarry exposures of the Malverns Complex. In most, there is a complex of chaotically sheared and altered lithologies. Figure 3 shows the location of the quarries.

1. North Malvern (Pyx) Quarries [768 469; 770 468]. The northerly quarry, also known as Tank Quarry, exposes mainly hybrid diorites, with some granites. The rocks are chaotically foliated and sheared, and zones of intense shearing are abundant, with schlieren of amphibolite and biotite schist. Epidote is common in joints and veins. There are numerous pegmatite and aplitic veins. A north-north-west - trending dolerite dyke is present at the north end of the quarry, and there are smaller north-south microdiorite dykes (Lambert and Holland, 1971; Penn and French, 1971; Bullard, 1989).

The more southerly quarry consists mainly of diorites with much acid veining. Barytes has been recorded in fault zones (Penn and French, 1971; Bullard, 1975).

2. Westminster Quarry [7652 4608]. The northern part of the quarry, also known as Westminster Arms Quarry, is in granites, foliated to a greater or lesser extent. Towards the south is a mixed zone of granitic and dioritic rocks, and, in the extreme south, a faulted zone with schist. There are a few basaltic dykes (Penn and French, 1971, p.33; Bullard, 1989, p.25). These authors suggested that the schist may be metasedimentary, but there

is no evidence for this interpretation, the rock being a sheared, schistose diorite.

3. Dingle Quarry [7654 4567]. A 4 m-thick dolerite dyke forms a step in the quarry. Granite veins in diorite below are truncated by the dyke. The dyke locally has a chilled margin against an overlying granite, but has a tectonic contact with the underlying diorite. (Penn and French, 1971, pp.32, 33; Bullard, 1989, pp.28, 29).

4. Hayslad (Dogleg) Quarry [767 449]. The shape of this quarry corresponds to a worked out north-south microdiorite dyke. Epidote is abundant in shears in the granitic host rock (Penn and French, 1971, p.31).

5. Upper County Quarry [768 448]. This quarry contains the remnants of an unsheared, east-dipping dolerite dyke, most of it having been quarried away. The dyke contains enclaves of granite. The west wall consists of granite and hard green schist (Lambert and Holland, 1971, p.328; Penn and French, 1971, p.31).

6. Lower (main) County Quarry [7675 4470]. Mesocratic and melanocratic diorites, dolerites and pink acidic veins are intruded into granite (Penn and French, 1971; Pocock and Whitehead, 1947, Plate IIB; also BGS photo A6241).

7. Earnslaw Quarry [770 440] exposes a potassium-rich trachyte dyke forming an east–west step across the quarry floor in the south (Thorpe, 1971). A fault zone immediately to the north of the dyke separates diorite to the north from granite to the south. Dolerite dykes are present throughout (Bullard, 1975).

8. Lower Tollgate Quarry [7705 4415] exposes a wide range of fine- to coarse-grained rock types, including granite, diorite, amphibolite and hornblende-biotite rocks. Epidote is abundant in shears. Brammall (1940) described a biotite-hornblende pyroxenite in a neck between the two parts of the quarry. Highly sheared microdiorite dykes occur in well-jointed, lineated, planar-foliated granite (Lambert and Holland, 1971, pp.330–331; Penn and French, 1971, p.30).

9. Upper Tollgate Quarry [7695 4395]. Pink, massive, foliated granite is the main rock type, with east–west elongation of quartz and biotite. A shear plane dips about $45°$ to the south in the main face, displaced to the north by a thrust, and there are $60°$-dipping joints. Small amounts of garnet are recorded and a microdiorite dyke is present in the eastern wall (Lambert and Holland, 1971, pp.229, 230; Penn and French, 1971, 1971, pp.29, 30).

10. Gardiner's Quarry [7660 4210] exposes a complex of sheared diorites and granites with much granitic veining. A sheared metadolerite is present in the east face. A shear zone at the southern end of the quarry may be the Colwall Fault (Phipps and Reeve, 1969; Penn and French, 1971, p.25; Bullard, 1975).

11. Pyxty Quarry [7690 4120]. A complex of sheared, altered and hydrolysed diorites and granites, thinly interleaved with more leucocratic material in places, is cut by quartzofeldspathic veins. Chlorite is abundant in the

shears. Large poikilotopic hornblende crystals occur locally on the south side of the quarry (Penn and French, 1971, p.25). A faulted, brecciated dolerite dyke is present in the upper part of the western wall (Bullard, 1975).

Deformation of the Malverns Complex

Fitch et al. (1969) proposed that the complex had a complicated Precambrian metamorphic history. This includes at least one high-grade regional metamorphic event of garnet-amphibolite facies associated with intense migmatisation, soon after emplacement of the plutonic protolith, followed by several episodes of retrograde metamorphism. Garnetiferous rocks occur in several places, for example on Swinyard Hill (South Malverns), North Hill and Midsummer Hill, mostly associated with sheared diorite and granitic rocks (Lambert and Holland, 1971). The garnet is invariably pseudomorphed by chlorite, biotite, muscovite, epidote or plagioclase. According to Lambert and Holland (1971), biotite and chlorite schists derived from diorite by cataclasis usually contain the assemblage calcic oligoclase and epidote. Albite-epidote assemblages are rare. Actinolite and tremolite occur widely. Earlier reports of sillimanite were probably misidentifications of fibrous epidote (Lambert and Holland, 1971). Metamorphosed dioritic rocks containing granulated hornblende separated by sericitised oligoclase (c.An_{30}) with subsidiary chlorite, epidote, quartz and opaque oxides were described as 'epidiorite' by Brammall (Lambert and Holland, 1971)*.

The early phase of prograde regional metamorphism was accompanied by heterogeneous ductile shearing, generating a fabric of variable intensity, in part a consequence of ductility contrasts between different components of the complex. Some of the ductile shear zones affect the granitic protolith and have produced banded, quartz-rich rocks superficially resembling sedimentary rocks. In zones of high strain, schistose mylonitic fabrics developed. Locally, for example at Wyche Cutting, frictional melting during episodes of fault motion resulted in the injection of pseudotachylite sheets discordant to the planar fabrics, probably at relatively shallow (i.e. <5 km) crustal depths (Thorpe, 1987). Subsequently the pseudotachylite veins were devitrified and hydrothermally altered. The absolute age of the early ductile phase of deformation is poorly constrained at present. Fitch et al. (1969) recorded biotite and muscovite K-Ar ages in the range 614–610 Ma, best interpreted as cooling (minimum) ages. This event may be responsible for Pb-loss from some zircons (p.10). A later generation of epidote with carbonate in veins, cross-cutting the mylonitic foliation, is probably associated with later, greenschist facies retrogression of the complex.

* Ar-Ar isotope correlation ages of c.550 Ma for hornblende syenites have been published since this memoir went to press (Strachan et al., 1996). They indicate that the igneous crystallisation and metamorphic recrystallisation of the complex represent temporally distinct events. These authors also recognise a subsequent phase of thermal reactivation, reflected in pegmatite emplacement and hydrothermal alteration at c.610–600 Ma.

Potassic and sodic pegmatites are discordant to the penetrative schistose and mylonitic fabrics in the protolith, but have themselves suffered deformation, as can be seen at Gullet Quarry [7618 3807] and Tank Quarry [7682 4706]. Lambert and Rex (1966) and Fitch et al. (1969) recorded biotite and muscovite K-Ar (minimum) ages of about 600 Ma for these pegmatites. The pegmatites are themselves cut by mafic dykes of granophyric diorite and microdiorite. Also, they are deformed by conjugate, late, brittle fractures which dip between 25° and 30°, as seen in Gullet Quarry. These fractures are probably associated with Palaeozoic inversion of the Malverns Complex.

Geochemical composition

The new geochemical analyses presented in Appendix 2 augment those of Blyth and Lambert (1970), Lambert and Holland (1971) and Thorpe (1974), which are largely major element data with a small number of trace elements. They are of 36 samples, collected from quarries in the Worcester and Tewkesbury districts. Data for another seven samples are previously unpublished results supplied by Dr T S Brewer. All of the new samples were analysed by X-ray fluorescence spectrometry at Nottingham University.

Petrogenetic studies of the plutonic rocks are hampered by the complexity of the magmatic and metamorphic processes which have affected the suite. There is abundant field evidence for hybridisation and metasomatism (Blyth and Lambert, 1970; Lambert and Holland, 1971; Thorpe, 1974). Subsequent alteration and metamorphism may have modified the primary magmatic composition still further, particularly the relatively mobile elements such as K_2O, Na_2O and CaO. The observed mineralogy of the suite is very different from calculated CIPW normative values due largely to the development of hornblende at the expense of normative feldspar, pyroxene and olivine (Dearnley, 1990). No attempt is therefore made here to classify the rocks using the IUGS scheme (Streckeisen, 1967). The rocks of the Malverns Complex are divided into six groups on their geochemical characteristics.

CUMULATE AMPHIBOLITES

Ultramafic and mafic amphibolites with high content of MgO (>12 wt%), Cr (>300 ppm) and Ni (>100 ppm) and generally low content of high field strength (HFS) elements such as Ti, Nb, Y and Zr are considered to result from the accumulation of ferromagnesian phases such as olivine or orthopyroxene, although this can rarely be confirmed petrographically. Data for these samples are presented in Figure 4a, in the form of geochemical patterns in which the data are normalised with respect to mid-ocean ridge basalt (MORB) values (Pearce, 1982). The most striking feature of the patterns is the wide spread of values exhibited by large ion lithophile (LIL) elements such as K, Rb and Ba, which have clearly been affected by alteration processes. By contrast, HFS elements such as Ti, Zr and Y appear less affected, and are presumably less mobile. Greater reliance is therefore placed on the variation of the

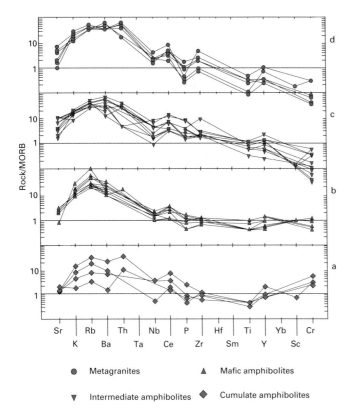

Figure 4 Geochemical patterns, with MORB-normalised values after Pearce (1982).

a Cumulate ultramafic and mafic amphibolites
b Non-cumulative mafic amphibolites
c Amphibolites of dioritic and tonalitic composition
d Metagranitic rocks.

HFS elements in the interpretations which follow. These observations are compatible with those of Lambert and Holland (1971), who recognised that the ultramafic amphibolites, particularly biotite-bearing varieties, had suffered the most severe hybridisation and metasomatic alteration of any component of the complex. Two samples (E58232 and E58233) contain felsic veins in hand-specimen, and their analyses are not depicted.

MAFIC AMPHIBOLITES

Data for amphibolites of gabbroic and mafic dioritic composition (with MgO+CaO>12 wt%), lacking geochemical evidence for cumulation, are shown in Figure 4b. Compared to the cumulate amphibolites, the geochemical patterns exhibit far less scatter, reflecting a lesser degree of alteration. The patterns in the HFS elements from P to Y are flat, with concentrations slightly less than MORB. Ce exhibits a significant enrichment with respect to Nb, indicating a subduction magmatic component (Pearce, 1982) in the profiles. The volcanic arc affinities of the mafic amphibolites are confirmed by the Ti-Zr-Y diagram (Pearce and Cann, 1973) and Nb-Zr-Y diagram (Meschede, 1986), in which the amphibolites plot in the fields of MORB and calc-alkaline arc magmatic rocks (Figures 5a and 5b).

INTERMEDIATE AMPHIBOLITES

Amphibolites of dioritic and tonalitic composition (Figure 4c) show strong enrichment in Ce, P and Zr, with respect to Ti and Y, so that the right hand side of the geochemical pattern slopes downward from Zr to Y, in contrast to the flat profile shown by the mafic amphibolites in this part of their patterns. This suggests that the intermediate rocks are not derived from the more mafic compositions simply by the fractionation of olivine and pyroxene. Either fractionation has involved a phase containing Y (e.g. garnet or hornblende), perhaps as a residuum from partial-melting, or a more complex process, such as one involving crustal contamination, has played a part. The depletion of Nb with respect to Ce and Th, observed in the mafic compositions, is retained. This suggests that the intermediate amphibolites may have been derived from a subduction-modified mantle source similar to that of the mafic amphibolites. Two samples (E58259 and E58261) exhibit macroscopic evidence of severe hydrothermal (argillic) alteration and their analyses are not depicted.

METAGRANITIC ROCKS

The MORB-normalised geochemical patterns of metamorphosed granitic rocks are depicted in Figure 4d. The overall shape of the patterns mirrors that of the intermediate amphibolites, particularly those with tonalitic compositions. Depletions of P and Ti with respect to the adjacent elements reflects fractionation of these elements in phases such as apatite and ilmenite. To facilitate comparison with other felsic components of the complex, the data for the granites are normalised with respect to ocean-ridge granite (ORG) values (from Pearce et al., 1984) in Figure 6. The most distinctive feature of the ORG-normalised patterns is the depletion in Nb which is found in many granites emplaced in volcanic arc and syn-collisional environments (Pearce et al., 1984, Fig.1). Figure 7, the Nb-Y diagram of Pearce et al. (1984), confirms the affinities of the Malverns Complex granites with those emplaced in volcanic arc and syn-collisional environments. Two felsic mylonites (E58255 and E58268) are believed, on petrographic and geochemical criteria, to have been derived from a granitic protolith.

PEGMATITES

Only limited trace element data are available for the pegmatites (Figure 6b). Contents of Nb, Zr, Ti and Y are all very low. The wide scatter of data on the normative Qz-Ab-Or diagram (Figure 8) indicates no simple relationship to the ternary eutectic, unlike the granites, which, as Thorpe (1974) has described, cluster around the isobaric minima.

TRACHYTE

Data for the small intrusive trachyte body in Earnslaw Quarry, depicted in Figure 6c, are those published by Thorpe (1971). The shape of the patterns is comparable to those of the granitic and intermediate rocks and the trachyte is clearly a cogenetic component of the Malverns Complex.

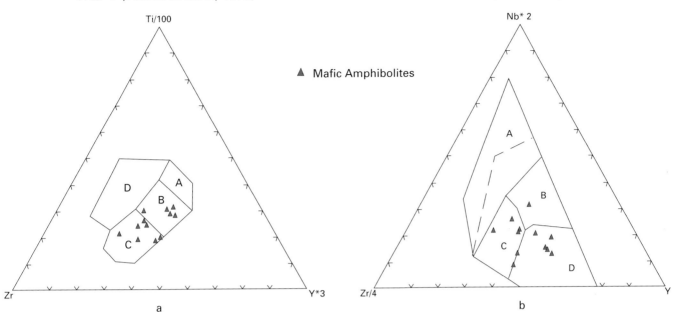

Ti-Zr-Y (Pearce & Cann, 1973)   Nb-Zr-Y (Meschede, 1986)

▲ Mafic Amphibolites

a   b

Figure 5

a Ti-Zr-Y variation diagram (Pearce and Cann, 1973) for mafic amphibolites (MgO+CaO wt%)
   of the Malverns Complex. Fields: A — island arc; A+B+C — MORB; C — calc-alkaline arc;
   D — within-plate
b Zr-Nb-Y variation diagram (Meschede, 1986) for mafic amphibolites of the Malverns
   Complex. Fields: A — within-plate; B — P-type MORB; D — N-type MORB;
   C+D — volcanic arc basalts.

Petrogenetic history of the Malverns Complex

The complex comprises a geochemically varied suite of rock types characterised by low content of Nb and Ti, lack of Fe-enrichment with magmatic evolution, as shown by the AFM diagram (Figure 9), and strong enrichment of certain LIL elements and Th. These are characteristics of calc-alkaline arc magmatic suites. This supports the subduction-zone origin for the complex proposed by Thorpe (1972, 1974) and Thorpe et al. (1984). Some features of the geochemical data, such as the divergent geochemical patterns exhibited by the intermediate and felsic compositions compared to the mafic compositions, suggest that the former are probably not derived simply by greater degree of crystal fractionation of olivine and/or pyroxene from the same parental liquid as the mafic components. An origin by heterogeneous partial melting of a mafic source and/or widespread crustal contamination of mafic magma are likely possibilities.

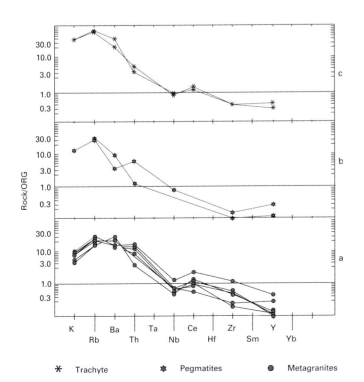

Figure 6 Geochemical patterns, with ORG-normalised values after Pearce et al. (1984).

a Metagranitic rocks
b Pegmatites
c Trachytes (data from Thorpe, 1971).

✳ Trachyte   ✴ Pegmatites   ● Metagranites

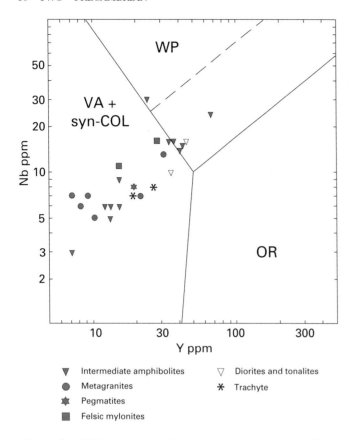

Figure 7 Nb-Y variation diagram for metagranites of the Malverns Complex. Fields after Pearce et al.(1984).

OR — ocean ridge; WP — within-plate; VA — volcanic arc; syn-col — syn-collision.

Age of the Malverns Complex

Isotopic evidence indicates that the Malverns Complex forms one of the oldest components of the Avalonian accretionary terrane of southern Britain. Only the Stanner-Hanter Complex, near Kington, 55 km west of the Malverns, which has yielded an Rb-Sr isochron age of 702 ± 8 Ma (Patchett et al., 1980), is demonstrably older. Tucker and Pharaoh (1991) published U-Pb isotopic data for zircon fractions from a deformed tonalite in Tank Quarry [7691 4701]. Four zircon fractions (Figure 10, 1a, b, c, e) fall on a discordia line with intercept ages of 677 ± 2 Ma, interpreted as the age of emplacement, and 1598^{+32}_{-30} Ma, interpreted as the average age of an inherited Proterozoic zircon component. A fifth fraction (d) does not fit on the discordia line, apparently indicating slight secondary Pb-loss. Within the (2δ) analytical errors, the younger age agrees with a concordant U-Pb monazite age of 670 ± 10 Ma (Thorpe et al., 1984) and a Rb-Sr whole-rock isochron age of 681 ± 53 Ma (Beckinsale et al., 1981), also interpreted as the age of emplacement. The inherited zircon fraction does not necessitate the presence of older crust beneath the Malverns, but could have been acquired by magmatic interaction with sedimentary rocks derived from older crust (Thorpe et al., 1984). The low initial

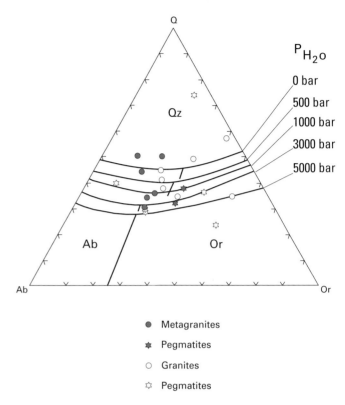

Figure 8 CIPW normative data for granites and pegmatites of the Malverns Complex plotted in the system Ab-Or-Silica-H_2O at a range of fluid pressures, after Tuttle and Bowen (1958).

$^{87}Sr/^{86}Sr$ ratio (0.705) precludes a prolonged crustal history.

PRECAMBRIAN MICRODIORITIC MINOR INTRUSIONS

Dykes of broadly mafic composition occur throughout the Malverns Complex, and are rather less deformed in the northern Malvern Hills than in the south (Worssam et al., 1989). According to Blyth and Lambert (1970), these mafic microdiorite sheets cut the pegmatites, and they certainly cross-cut and postdate development of the ductile fabric in the Malverns Complex. The most common type is a granophyric quartz microdiorite (frequently referred to as 'dolerite' in the literature). In thin-section, andesine laths are poikilitically enclosed in clinopyroxene or secondary hornblende, the latter commonly chloritised. Interstitial intergrowths of micrographic quartz and feldspar occur. The other common type is hornblende microdiorite, which has a non-ophitic texture, with highly zoned andesine and green hornblende showing alteration to chlorite. Both types of microdiorite intrusion exhibit well preserved igneous texture, pseudomorphed to varying degrees by greenschist facies assemblages (R J Merriman, personal communication, 1992). The microdiorite intrusions are presumed to be

Igneous AFM wt%

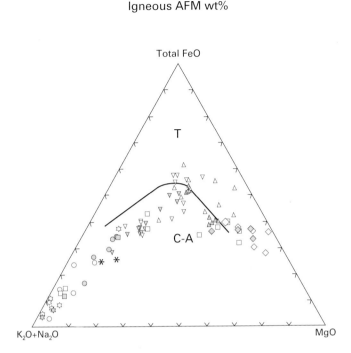

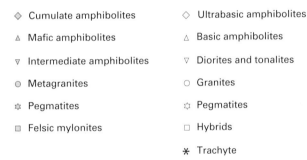

◇ Cumulate amphibolites   ◇ Ultrabasic amphibolites

△ Mafic amphibolites   △ Basic amphibolites

▽ Intermediate amphibolites   ▽ Diorites and tonalites

◉ Metagranites   ○ Granites

✿ Pegmatites   ✩ Pegmatites

▣ Felsic mylonites   ▢ Hybrids

                                ✳ Trachyte

Figure 9   AFM diagram for samples of the Malverns Complex; boundary between tholeiitic (T) and calc-alkaline (C-A) fields after Irvine and Barager (1971).

Geochemical composition

Four of the samples collected (Appendix 2, Table 3) exhibit textural and metamorphic evidence indicating affinity with the microdiorite suite, and not the Malverns Complex (R J Merriman, personal communication, 1992). Limited trace element data for these bodies were published by Lambert and Holland (1971) and Thorpe

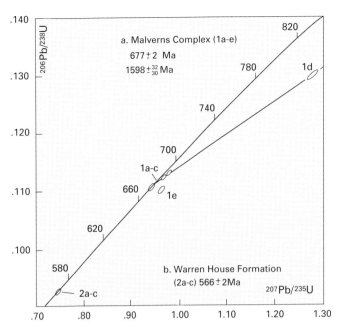

Figure 10   Concordia diagram, modified from Tucker and Pharaoh (1991), of U-Pb zircon analyses from:

a Deformed tonalite, Malverns Complex, North Malvern.
b Felsic tuff, Warren House Formation, Broad Down.

(1974). For comparison, data from the Ordovician spilitic microdiorite suite (Worssam et al., 1989) intruding Cambrian–Tremadoc strata in the south Malvern Hills at Hollybush and in the Fowlet Farm Borehole (Worssam et al., 1989) are also depicted in Figure 11.

The geochemical patterns (Figure 11a) are very similar to the published data for the granophyric microdiorite (Figure 11b) and hornblende microdiorite (Figure 11c), exhibiting strong enrichment of LIL elements and Th, moderate depletion of Nb, and a gently sloping profile in the HFS elements from Ce to Y, comparable to those of typical within-plate basalts (Pearce, 1982). The Precambrian microdiorite intrusions have higher contents of Zr, Y and Ti than the Ordovician intrusions (Figure 11d), which exhibit geochemical patterns similar to the mafic components of the Malverns Complex. The Ordovician suite exhibits strong dispersion of K and Rb (much less so for Ba and Th), which testifies to the severe, spilitic alteration recognised on petrographic criteria (Worssam et al., 1989). The strong Fe enrichment displayed by the Precambrian, mafic, microdiorite bodies is reflected in the FMA diagram (Figure 12), on which the majority fall within the tholeiitic field. By contrast, the Ordovician intrusions plot entirely within the calc-alkaline field of Figure 12.

Thorpe (1974) has emphasised the transitional tholeiitic to mildly alkaline character of the Precambrian, mafic, microdiorite bodies. They form a comagmatic suite clearly distinct from the Malverns Complex on geochemical (as well as structural and metamorphic) criteria. The enrichment in LIL elements and depletion of

of Precambrian age, as Cambrian sedimentary rocks in the south Malvern Hills have significantly lower grade, diagenetic zone mica crystallinity values (Pharaoh, unpublished data). The term 'Ivy Scar' type was applied first to the granophyric (Blyth and Lambert, 1970) and subsequently, to the hornblendic (Lambert and Holland, 1971) varieties. A sample from the Ivy Scar locality (E66190) does contain granophyric patches (R J Merriman, personal communication, 1992). In the light of this confusion, it is recommended that usage of the term 'Ivy Scar' type is abandoned.

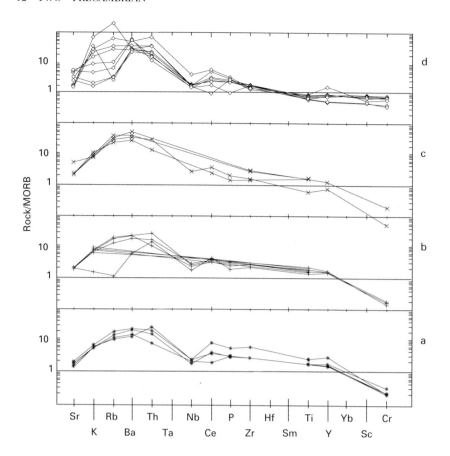

Rock/MORB

Post-Cambrian microdiorites ◇

Hornblende microdiorites ×

Granophyric microdiorites +

Microdiorite dykes *

Figure 11 Geochemical patterns, with MORB-normalised values after Pearce (1982).

a Mafic microdiorite intrusions
b Granophyric microdiorite (data from Thorpe, 1974)
c Hornblende microdiorite (data from Thorpe, 1974)
d Post-Cambrian (?Ordovician) spilitic microdiorite intrusive suite, south Malvern Hills.

Igneous AFM wt%

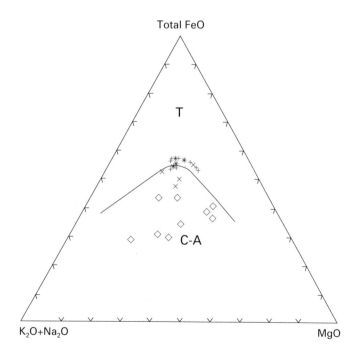

Total FeO

T

C-A

K₂O+Na₂O   MgO

* Microdiorite dykes   × Hornblende microdiorites

+ Granophyric microdiorites   ◇ Post-Cambrian microdiorites

Nb, noted above, may be an inheritance from a sublithospheric mantle region modified by subduction magmatism during the formation of the Malverns Complex. In terms of regional correlations, the mafic minor bodies most closely resemble basaltic-andesite lavas of the Uriconian volcanic group (Pharaoh et al., 1987) and are distinct in trace element composition from the lavas of the Warren House Formation.

WARREN HOUSE FORMATION

The Warren House Formation crops out from Tinker's Hill [767 401] south to Hangman's Hill [765 390] on the adjoining Tewkesbury sheet. The formation comprises spilitic basalt lavas, locally displaying pillow texture (e.g. Worssam et al., 1989, Plate 2), altered intermediate lavas ('keratophyres'), altered rhyolitic lavas and felsic pyroclastic rocks, including ignimbrite. The volcanic rocks are intruded by north-trending dolerite dykes, one of which is exposed at the reservoir quarry [766 397]. According to French and Winchester (personal communication to Lambert and Holland, 1971), the formation occurs in an

Figure 12 AFM diagram for mafic, minor bodies intruding the Malverns Complex. Boundary between tholeiitic (T) and calc-alkaline (C-A) fields after Irvine and Barager (1971).

overlain by felsic rocks to the east. Excavations at the reservoir [765 399] at the time of its construction (Acland, 1898; Green, 1895) revealed that the situation is more complex, with a sequence of interbedded basalts, rhyolites and tuffs and an easterly dip ranging from 45° to near vertical (Worssam et al., 1989).

The petrography of the various rock types was described by Platt (1933). The preservation of pillow textures and other primary textures, and the weakness of the greenschist facies foliation contrasts strongly with the much more intense deformation and metamorphism of the Malverns Complex. The contact between the two units is tectonic, and the tectonism must postdate eruption of the Warren House lavas at 566 Ma (see below). This is here attributed to late Avalonian tectonism at the end of the Proterozoic, possibly the same event during which the Longmyndian Supergroup was folded.

Geochemical composition

Thorpe (1972, 1974) studied the chemistry of the lavas and concluded that the heavily altered basalts were originally tholeiitic lavas of possible ocean-floor affinity, although the presence of felsic tuffs causes some problems with this interpretation. Pharaoh et al. (1987) studied the trace element geochemistry of the lavas and demonstrated a strong enrichment in large ion lithophile (LIL) elements such as Ba, and Th, whereas the content of high field strength (HFS) elements such as Ti, Y and Zr, lies close to or just below that of mid-ocean ridge basalts (MORB). These geochemical characteristics are most similar to those of basalts from primitive volcanic arc and marginal basin complexes in the western Pacific (Saunders and Tarney, 1984), and Pan-African ophiolites (proposed marginal basin crust) of the Arabian Shield (e.g. Klemenic, 1987). The geochemical patterns of the Warren House intermediate and felsic lavas parallel those of the basaltic rocks, and suggest that they may be derived from the latter by differentiation. Pharaoh et al. (1987) concluded that the Warren House lavas were probably erupted in an oceanic marginal basin and preserved as a tectonic sliver along a Proterozoic suture between the Wrekin and Charnwood terranes.

Age of the Warren House Formation

A Precambrian age, inferred from the presence of probable Warren House Formation pebbles in the Lower Cambrian Malvern Quartzite (Earp and Hains, 1971), has recently been confirmed by radiometric dating using the U-Pb isotope system. Zircon fractions from a crystal-lithic tuff of rhyolitic composition exposed on Broad Down [7655 3960] yielded overlapping and concordant analyses (Figure 10), defining an eruption age for the tuff of 566±2 Ma (Tucker and Pharaoh, 1991). This age is identical to that obtained for the Uriconian volcanic group of Shropshire (Tucker and Pharaoh, 1991). The formation has been correlated with the Uriconian lavas (Callaway, 1880; Earp and Hains, 1971) on lithological criteria. However, there are geochemical contrasts between the two suites which suggest that the Warren House Formation and the Uriconian

formed in oceanic and ensialic marginal basin environments respectively (Pharaoh et al., 1987).

CONCEALED PRECAMBRIAN BASEMENT

Kempsey Formation

Up to 3000 m of Permo-Triassic strata overlie the basement in the middle of the Worcester Basin (Chapter 10). The Kempsey Borehole [8609 4933], drilled by the Survey with support from the Department of Energy, proved 706 m of Precambrian volcaniclastic rocks, here named the Kempsey Formation, beneath 2305 m of Permo-Triassic strata (Chapter 7; Whittaker, 1980). Precambrian rocks floor the area of the basin within the Worcester district (Smith, 1985), Palaeozoic sedimentary rocks having been eroded subsequent to Variscan (late-Carboniferous) inversion (Chadwick and Smith, 1988; Chapter 10).

The formation comprises lithic sandstone and tuffaceous siltstone, purplish green in colour, with poorly sorted angular to subrounded clasts of tuff, agglomerate and lava of acid to intermediate composition. Cored samples were recovered at depths of 2336.0 to 2338.4 m, 2459.1 to 2462.8 m and 3009.0 to 3011.4 m. At 2336 m, poorly sorted agglomerates and tuffs contain fragments of greyish green and buff welded and crystal tuff, white to pink siliceous felsite, green and red very fine-grained tuff, up to 3 cm across, and sand-grade quartz fragments. Many of these fragments have marginal reaction zones or hematitic staining, indicative of penecontemporaneous weathering. Petrographic studies by R K Harrison indicate that the clasts include microagglomerate of (reworked) trachyandesite and trachyte, felsite, devitrified rhyolite, chloritised glass and pumice. At 2459 m, bedded tuffs contain fragments of tuffaceous material up to 8 cm across. Some beds exhibit fining-upwards grading, cross-bedding and erosive channel-like features. The fragments range from angular (volcanic clasts) to rounded (sand grains) and include vitric tuff and heavily altered fragments of more basic lava (andesite or spilitic basalt) in addition to the types described previously. At 2983 m, rather massive reddish brown tuffaceous sandstones are developed. The apparent dip of bedding is moderately steep (20–60°) throughout the sequence. Veins of quartz and calcite, hematitic impregnations and epidotic matrix are ubiquitous.

GEOCHEMICAL EVIDENCE

Major and trace element geochemical data for core samples from the Kempsey Borehole are listed in Appendix 2, Table 4. Four clasts of grey-green lava from 2336.0 to 2338.4 m are of intermediate composition. The most obvious feature of the geochemical patterns is the strong enrichment in LIL elements such as K, Rb and Ba, Th, P and light REE (e.g. Ce) with respect to HFS incompatible elements such as Nb, Zr, Ti and Y. The content of HFS elements is low, less than that for MORB, indicating that the mantle source for the lavas was relatively depleted in these elements, like that of the

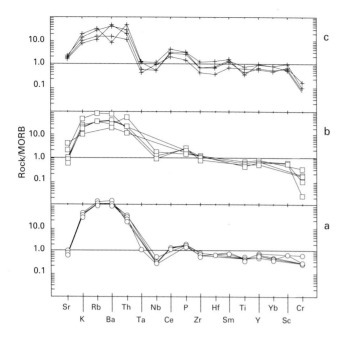

Figure 13 Geochemical patterns, with MORB-normalised values after Pearce (1982) for:

a   Clasts of intermediate tuff from core samples of the Kempsey Formation

b   Tuffaceous matrix to clasts and lithic sandstones from core samples of the Kempsey Formation

c   Diorites from the Charnwood area, representative of the Charnian magma type (from Pharaoh et al., 1987).

Warren House Formation and Charnian magmas, and unlike that of the Uriconian magmas which have a strong within-plate component (Pharaoh et al., 1987). The degree of enrichment in LIL and REE in the Kempsey Formation is most comparable to that of the Charnian 'magma-type'. Geochemical data from tuffaceous matrix from 2338 m, and lithic sandstones from 2461 m and 3009.6 m, are plotted in Figure 13b. These fragmental samples are slightly more mafic than the lava clasts, but exhibit very similar geochemical patterns, and are clearly derived from the same volcanic arc source as the lava clasts.

AGE AND AFFINITY OF THE KEMPSEY FORMATION

No radiometric data are currently available for the age of the formation. The geochemical data indicate that the magmatic source of the volcaniclastic material could have been the same as that of the Charnian Supergroup. As the volcaniclastic rocks are demonstrably reworked, it is conceivable that Precambrian fragments could have

been reworked into younger (i.e. Palaeozoic) sedimentary rocks. However, the following observations suggest that such an interpretation is unlikely. Firstly, the compositions of fine-grained tuffs and lithic sandstones at various depths, and of the lava clasts, are very similar, and suggest that one magmatic source remained dominant throughout the accumulation of the 706 m of beds proved, and that the volcaniclastic sedimentary rocks are unlikely to be polycyclic. Secondly, the top of the formation in the Kempsey Borehole is coincident with a bright reflection event on seismic profiles across the Worcester Basin, interpreted as the top of the 'crystalline basement' (Chadwick, 1985; Chadwick and Smith, 1988); the latter shows no sign of the internal reflectivity characteristic of Palaeozoic sequences. Thirdly, borehole geophysical logs indicate that the Kempsey Formation is a monotonous succession of interbedded fine- to coarse-grained sedimentary rocks, comparable in lithology (by gamma log), density ($2.75\ Mg^{-3}$) and sonic velocity ($5750\ ms^{-1}$) to Charnian volcaniclastic sedimentary rocks of the type area of Charnwood (Pharaoh and Evans, 1987).

Pharaoh et al. (1991) have inferred that the formation is contemporaneous with the Charnian Supergroup, and was derived from the same volcanic arc source. On this basis, they infer that the Charnwood Terrane (Pharaoh et al., 1987) extends as far west as the East Malvern Fault. The Precambrian Worcester and Charnwood marginal basins are separated by a basement ridge, running from Birmingham to Reading, associated with a major, deep-sourced aeromagnetic anomaly. Lee et al. (1991) have suggested that this may mark the magmatic core of the Charnian (Precambrian) volcanic arc, although a younger, Carboniferous age has also been proposed for the magnetic body causing the anomaly (Kearey, 1991).

West of the Malvern Axis

To the west of the Malvern Axis, a major monoclinal fold takes the Precambrian basement down to a depth of 1500 to 2000 m beneath the western part of the district (Chapter 10). Information on the nature of the Precambrian basement in this area is sparse and comes principally from geophysical evidence. The exploratory hydrocarbon Collington Borehole [6460 6100] terminated in a rock type yielding fragments of altered alkali-rich granite at 1710 m depth. It is not certain that the borehole proved basement, and the granite fragments may have been derived from clasts in Silurian strata. Unfortunately, no samples have been preserved to compare with Malverns Complex lithologies. The nature of the Precambrian basement to the Lower Palaeozoic rocks to the south is discussed in Chapter 10.

THREE

Cambrian and Ordovician

Cambrian rocks have been proved at only four localities in the district and are poorly represented in comparison to their occurrence at the southern end of the Malverns, where the sequence (Cowie, 1992; Worssam et al., 1989) comprises:

Bronsil Shale, silver-grey mudstones	250–350 m
White-leaved Oak Shale, dark grey mudstones	150–250 m
Hollybush Sandstone, dark green glauconitic sandstones	300 m
Malvern Quartzite, pale grey sandstones and conglomerates up to	about 60 m

The Bronsil Shale, of Tremadoc age, is now placed in the Ordovician. All the formations comprise the Dyfed Supergroup of Woodcock (1990). Their presence at depth in the south of the district is inferred from their occurrence at outcrop in the southern Malverns. Only the Malvern Quartzite and White-leaved Oak Shale are present at outcrop or in the shallow subsurface in the Worcester district.

A small roadside exposure [7634 4765] of shattered white quartzite occurs at Cowleigh Park in a fault zone comprising mainly Malverns Complex granite and Llandovery sedimentary rocks (Figure 14; Phillips, 1848; Groom, 1900, fig.11). It is presumed to be a remnant of the Malvern Quartzite. Similar quartzite occurs in thrust contact with the Malverns Complex at Martley to the north of the district (Mitchell et al., 1962).

The record of a well [7654 4734] at West Malvern (Mackie, 1887; Groom,1900; Ziegler, Cocks and McKerrow, 1968; Brooks, 1970; Bullard, 1974) includes a description of beds attributed to the White-leaved Oak Shale (Bed 5), faulted between Llandovery strata. The section was described (Mackie, 1887) as:

	Thickness
1. Surface gravel (hill debris)	3.66 m (12 ft)
2. Red crumbling marl with small pebbles	7.32 m (24 ft)
3. Coarse grey and red sandstone and conglomerate	4.88 m (16 ft)
4. Very hard breccia of quartzite pebbles in a red and yellow matrix	1.83 m (6 ft)
5. Black Cambrian shale with *Olenus*, *Conocoryphe*, *Lingulella*,etc.	5.49 m (18 ft)
6. Repetition of beds 2 and 3	3.96 m (13 ft)

Dr A W A Rushton comments that the fossils recorded from the White-leaved Oak Shale do not give a clear indication of age. The trilobite genera *Olenus* and *Conocoryphe* were both formerly used in a wide sense that en-

compassed species occurring together in the Merioneth and Tremadoc series. Re-examination of the fossils is necessary, but although they were said to have been 'sent to Cambridge' (Groom, 1900, p.158), none have been traced.

Brooks (1970) considered that there was a considerable thickness of Cambrian beds below the Llandovery sequence, which were cut out to the east by pre-Llandovery movement on the reversed fault that forms the western boundary of the Malverns Complex at this locality. Boreholes and temporary sections at North Malvern in 1980 (p.25) mostly showed Llandovery beds in unconformable or faulted contact with the Malverns Complex. One borehole proved 0.8 m of dark grey to black silty clay and brecciated black shale above granite. There are thus only isolated thin remnants of Cambrian strata in this vicinity. It is possible that they thicken westwards from here, away from the Malverns, as suggested by Brooks (1970) and Ziegler, Cocks and McKerrow (1968). This may be the case on the southern margin of the district, but in the light of recent seismic evidence and of the record of the Collington Borehole (Chapter 10), it is unlikely that Cambrian–Ordovician strata extend far north of the district's southern boundary.

Bullard (1974) recorded 4 cm of black, fine-grained clayey shale with two very small lingulids at a temporary exposure [7655 4730] in West Malvern. He considered that this may be either a faulted remnant of Cambrian strata adhering to the Precambrian surface, or the basal bed of the Llandovery. His structural interpretation, implying a thick sequence of Cambrian strata in the core of a faulted anticline to the west, is unlikely.

By analogy with the sequence to the south, the Malvern Quartzite and Hollybush Sandstone are assigned a Lower Cambrian, Comley Series age, and the White-leaved Oak Shale an Upper Cambrian, Merioneth Series age, with the St David's Series unrepresented (Cowie et al., 1972; Groom, 1902; Worssam et al., 1989).

Little of the depositional and tectonic history of the Cambrian can be deduced from the few occurrences of Cambrian strata within the district. From evidence elsewhere, the accretionary phase of the Late Precambrian culminated in inversion on the north-west margin of the Avalonian terrane, succeeded by extensional rifting and inception of the Welsh Basin (e.g. Brasier et al., 1992). Pebbles of Precambrian igneous and volcanic rocks in the Malvern Quartzite indicate erosion of the adjacent landmass and deposition in shallow marine waters on the south-eastern flanks of the Welsh Basin, onlapping eastwards onto the Midland Platform. The Malvern Axis may have been active in the Middle Cambrian, with the St David's Series absent. Euxinic black shale deposition of the White-leaved Oak Shale in the Upper Cambrian ex-

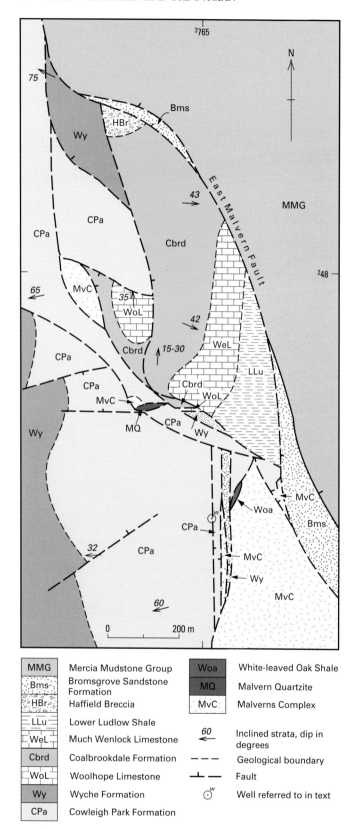

tended westwards into the district, but the extent of the transgression is not known, most of the deposits having been removed during Ordovician inversion. In the most recent reconstruction, Brasier et al. (1992) postulate that the margin of the depositional area may not have been far west of the Malverns, with a north-trending ridge separating the Malvern area from the Welsh Basin. Late Cambrian diastrophism was succeeded by more extensive drowning of the Midland Platform and deposition of the Tremadoc Bronsil Shales.

MMG	Mercia Mudstone Group	Woa	White-leaved Oak Shale
Bms	Bromsgrove Sandstone Formation	MQ	Malvern Quartzite
HBr	Haffield Breccia	MvC	Malverns Complex
LLu	Lower Ludlow Shale		
WeL	Much Wenlock Limestone	60 ←	Inclined strata, dip in degrees
Cbrd	Coalbrookdale Formation	– – –	Geological boundary
WoL	Woolhope Limestone		Fault
Wy	Wyche Formation	⊙ʷ	Well referred to in text
CPa	Cowleigh Park Formation		

Figure 14 Sketch map of the geology of the Cowleigh Park area.

FOUR

Silurian

Silurian rocks crop out west of the Malvern Hills in the south of the district, and on the Malvern Axis to the north. The oldest exposed strata are of mid- to late Llandovery Series age. These rest unconformably on, or are in faulted contact with, the Precambrian, except where Cambrian strata occur locally (p.15). The youngest rocks, the Raglan Mudstone Formation, belong to the Přídolí Series and are of continental Old Red Sandstone red-bed facies which spans the Silurian-Devonian boundary (p.36). Figure 15 shows a generalised vertical section of the sequence at outcrop and a section of the sequence proved in the Collington Borehole, situated close to the north-west of the district. All the beds are grouped into the Powys Supergroup of Woodcock (1990).

Most of the Llandovery, Wenlock and Ludlow rocks are of shallow-marine origin. Siliciclastic shelf sequences alternate with shallow carbonates with variable amounts of terriginous material. All were deposited on the southeast margin of the Lower Palaeozoic Welsh Basin and the western flanks of the Midland Platform. The Přídolí rocks are mainly alluvial coastal plain deposits, with minor amounts of shallow-marine to brackish-water intertidal deposits.

The rocks have been the subject of much lithostratigraphical, biostratigraphical, sedimentological and palaeoecological research, and included in many regional studies of the Welsh Borderland. First classified and described in some detail by Murchison (e.g. 1839), the sequence was mapped by Phillips during the first geological survey, described by him in the survey memoir published in 1848, and the one-inch map published in 1855. Since then, the sequence has been remapped and described in detail by Groom (1899, 1900, 1902), Phipps and Reeve (1967) and Penn (1969). Palaeoecological studies of the Llandovery benthic faunas have been made by Ziegler (1965), Ziegler, Cocks and Bambach (1968) and Ziegler, Cocks and McKerrow (1968), of the Wenlock and Ludlow by Calef and Hancock (1974), of the Wenlock by Hurst (1975a) and of the Ludlow by Watkins (1979) and Watkins and Aithie (1980). Wenlock stratigraphy and palaeogeography have been reviewed by Bassett (1974) and Hurst et al.(1978). In addition, the nature of the Precambrian–Llandovery boundary generated much discussion in the 1960s (see p.21 for references).

No systematic fossil collections were made during the present survey, but Dr D E White advised on selected localities where faunal identifications were required to solve mapping problems. The mapping carried out on the outcrop of the Raglan Mudstone Formation (and on that of the overlying Devonian rocks) has provided for the first time a detailed picture of that area.

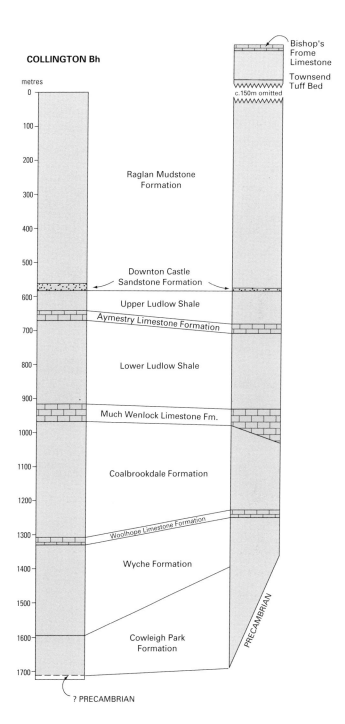

Figure 15 Section of Collington Borehole [6460 6100] and generalised vertical section of the Silurian rocks at outcrop; see p.14 concerning rocks of possible Precambrian age at base of Collington Borehole.

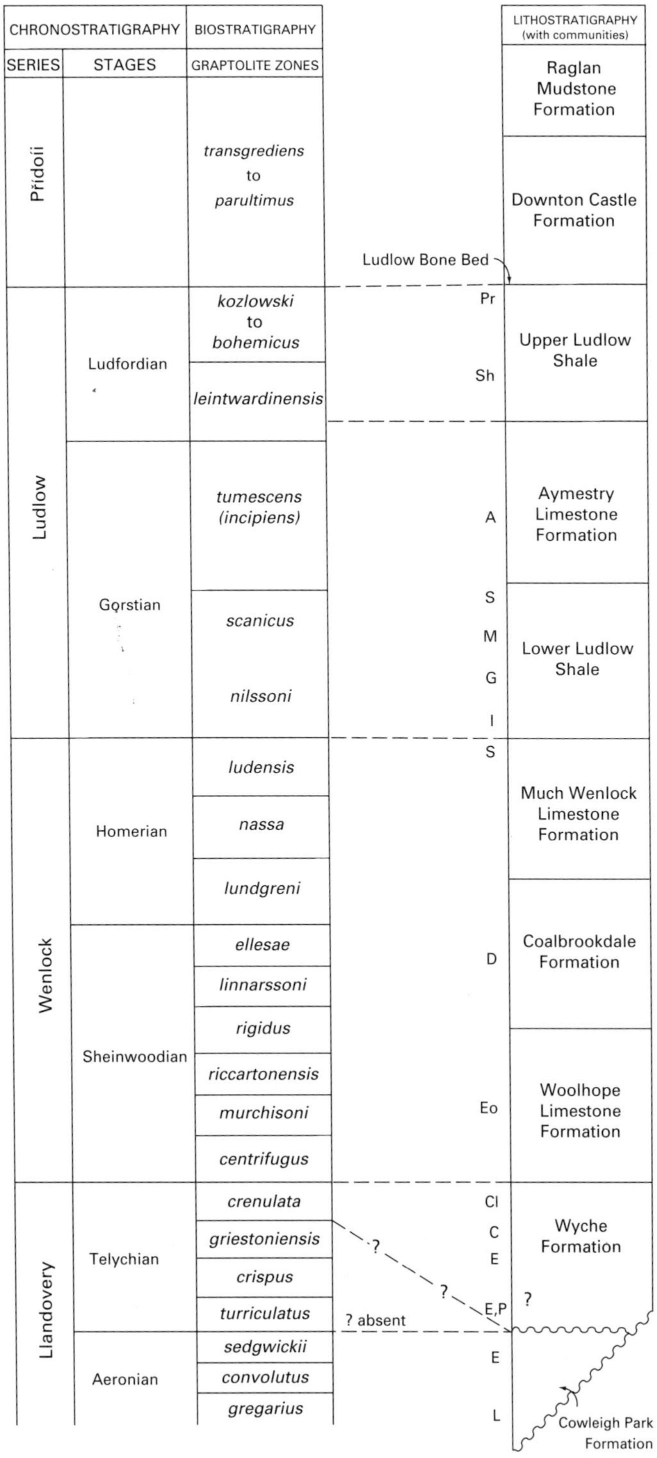

CHRONOSTRATIGRAPHY		BIOSTRATIGRAPHY	LITHOSTRATIGRAPHY (with communities)	
SERIES	STAGES	GRAPTOLITE ZONES		
Přídolí		transgrediens to parultimus	Raglan Mudstone Formation	
			Downton Castle Formation	
			Ludlow Bone Bed	
Ludlow	Ludfordian	kozlowski to bohemicus	Upper Ludlow Shale	Pr Sh
		leintwardinensis		
	Gorstian	tumescens (incipiens)	Aymestry Limestone Formation	A
		scanicus	Lower Ludlow Shale	S M G I
		nilssoni		
Wenlock	Homerian	ludensis	Much Wenlock Limestone Formation	S
		nassa		
		lundgreni		
	Sheinwoodian	ellesae	Coalbrookdale Formation	D
		linnarssoni		
		rigidus		
		riccartonensis	Woolhope Limestone Formation	Eo
		murchisoni		
		centrifugus		
Llandovery	Telychian	crenulata	Wyche Formation	Cl C E E,P
		griestoniensis		?
		crispus		
		turriculatus		
	Aeronian	sedgwickii	Cowleigh Park Formation	E L
		convolutus		
		gregarius		

? absent

Figure 16 Silurian stratigraphy.

Brachiopod communities/associations:
L *Lingula*; E *Eocoelia*; P *Pentameroides*; C *Costistricklandia*;
Cl *Clorinda*; Eo *Eoplectodonta duvalii*; D *Dicoelosia*;
S *Sphaerirynchia wilsoni*; I *Isorthis clivosa*; G *Glassia obovata*;
M *Mesopholidostrophia laevigata*; A *Atrypa reticularis* coral;
Sh *Shaleria ornatella*; Pr *Protochonetes ludloviensis*. Based on Calef
and Hancock (1974), Hurst (1975a), Watkins (1979) and
Ziegler, Cocks and McKerrow (1968).

CLASSIFICATION

The chronostratigraphical, biostratigraphical and lithostratigraphical classifications of the Silurian rocks are shown in Figure 16.

The Llandovery Series lithostratigraphical names are those of Groom (1910), given formation status by Phipps and Reeve (1967). The current Ordnance Survey spelling of Wyche (formerly Wych) is used for the Wyche Formation. For the Wenlock Series, the names proposed by Bassett et al. (1975) for the type area of Wenlock Edge, Shropshire are used, apart from the Buildwas Formation, for which the name Woolhope Limestone is retained. The classification of the Ludlow Series rocks of the type area immediately west of Ludlow, Shropshire has generated much debate (Holland et al., 1959, 1962, 1963; Phipps, 1962; 1963; Dorning, 1982; Lawson, 1982). This has centred mainly on the validity of the lithostratigraphical classification of the Ludlow Series in the type area, as proposed by Holland et al., and its application to other areas. Phipps criticised the units of Holland et al., which are based on a combination of lithostratigraphical and biostratigraphical criteria. He argued that lithostratigraphical classification is not based on contained faunas, but on gross mappable lithology, and should be discrete and separate from a biostratigraphical classification based on contained faunas. Holland et al. (1963) maintained that there are subtle lithological changes at the boundaries of the units in the type area, and Holland et al. (1980) considered them to be acceptable as mappable formations and renamed them as such. The units were retained as formations by Lawson and White (1989) in the type area and combined into groups by Siveter et al. (1989). Faunas characteristic of most of the units can be recognised in the Malverns (Phipps and Reeve, 1967; Penn, 1971, 1969; Cocks et al., 1992), but only the traditional broad lithostratigraphical tripartite subdivision of the Ludlow Series has proved practical to map in this district, following the practice adopted by the Geological Survey in the Church Stretton (Greig et al., 1968), Hereford (Brandon, 1989) and Tewkesbury (White et al., 1984; Worssam et al., 1989) districts. Table 2 shows this classification and its correlation with the others used in this district and in the Ludlow area. For the Přídolí Series, the name Raglan Mudstone Formation is retained to give continuity with the areas recently mapped to the south and west, rather than the name Ledbury Formation proposed by White and Lawson (1989).

FAUNA AND FLORA

The Llandovery, Wenlock and Ludlow rocks contain a diverse marine shelly macrofauna (Plates 1 and 2). Graptolites, however, are rare, and the biostratigraphical classification of the Silurian based on graptolites cannot be directly applied to the local succession. Of the shelly macrofossils, brachiopods have the most stratigraphical value, but ostracods are useful at some levels and may eventually form the basis of a biostratigraphical classification. Conodonts and acritarchs also provide biostratigraphical classifications. The Přídolí Series locally contains fish remains and ostracods which can be used for national and international correlation.

Particular associations of marine shelly benthos, mainly brachiopods, have been recognised throughout the Llandovery, Wenlock and Ludlow series. The development of these associations or communities depended on local conditions prevailing at the sea–seafloor interface. These conditions, including type of sea-bed sediment, water temperature, food supply and sedimentation rates, were largely, though not exclusively, depth-related, so that the distribution of fossil communities is useful in palaeogeographical reconstruction.

DEPOSITIONAL HISTORY

Global sea-level rise caused by late Ordovician deglaciation resulted in an eastward and southward extension of the Welsh Basin in Rhuddanian (earliest Llandovery) time. By late Aeronian time the sea had reached the Malverns, and the earliest Llandovery rocks of the district, the Cowleigh Park Formation, were deposited to the west of a shoreline that lay along the Malvern Axis. A regressive event from latest Aeronian to mid-Telychian times, indicated by a non-sequence in the contained faunas (Ziegler, Cocks and McKerrow, 1968; Cocks, 1989), led to deposition of continental red beds at the top of the Cowleigh Park Formation. Renewed transgression produced onlap of the Cowleigh Park Formation by the Wyche Formation, which, together with records of successively deeper-water *Lingula* to *Clorinda* communities, indicates a progressively deepening sea.

The oldest Wenlock (early Sheinwoodian) rocks are the carbonates of the Woolhope Limestone, which contain an *Eoplectodonta duvalii* Community (Hurst, 1975a). This carbonate facies community is considered to be the equivalent of both the Llandovery *Clorinda* and the Wenlock *Dicoelosia* siliciclastic facies communities. The latter extends through the Coalbrookdale Formation, suggesting unchanged bottom conditions from the late Telychian into the early Homerian. Evidence of regression during the late Homerian is provided by the Much Wenlock Limestone, which contains small reef bodies. These bodies have been compared to modern patch reefs, developing on the sea floor at less than 30 m depth, below wave base (Scoffin, 1971). As in the type area of Wenlock Edge (Scoffin, 1971), there is evidence in the Malvern area of extreme shallowing at the close of the Wenlock Epoch. Hurst (1975b) interpreted the topmost beds of the limestone at Whitman's Hill [743 483] as having formed above wave base, with shallow neptunean dykes in the uppermost 0.5 m. During the subsequent transgression, these fissures were filled with crinoidal limestone debris and green shale fragments closely comparable to the overlying Lower Ludlow Shale. At the top of the limestone, Hurst reported a shallow-water, nearshore *Sphaerirhynchia wilsoni* Community, replaced in the overlying Lower Ludlow Shale by the deeper-water *Isorthis*

clivosa and *Dicoelosia* communities at Whitman's Hill and at Park Wood [761 442].

During the Ludlow Epoch, deposition in the Welsh Borderland gradually changed from distal to proximal shelf as shoreline progradation took place north-westwards into the basin. This is reflected in the amount of bioturbation throughout, an increase in sediment grain size and in sedimentation rates, increasing amounts of shell transport and the frequency and intensity of storm events (Watkins, 1979, pp.251, 252). The change is also indicated by the sequence of benthic communities in the Lower Ludlow Shale. In the lower part of the formation, the *Dicoelosia* Community of Calef and Hancock (1974), which is the equivalent of the *Glassia obovata* Association of Watkins (1979), is succeeded through a transitional assemblage by the *Mesopholidostrophia laevigata* Association and by the *Sphaeriryhynchia wilsoni* Association at the top.

The Aymestry Limestone Formation, of latest Gorstian age, represents an interruption to shoaling siliciclastic deposition. It consists of intensely bioturbated silty and muddy limestones deposited in a low-energy back-barrier setting, protected by *Kirkidium*-rich banks on the shelf edge which are now exposed in the Aymestrey and Leintwardine areas (Watkins and Aithie, 1980). The restricted environment probably accounts for the paucity of the macrofauna which belongs to the *Atrypa reticularis*–coral Association of Watkins (1979).

Within the transitional beds at the base of the Upper Ludlow Shale, of earliest Ludfordian age, beds of limestone conglomerate have been recorded at numerous localities in the Welsh Borderland, including Chances Pitch [746 402] (Cherns, 1980) and Halesend [738 487] (Penn, 1969). Bored pebbles in the conglomerates indicate hardground formation during pauses in sedimentation (Cherns, 1980).

The succeeding beds of the Upper Ludlow Shale indicate a return to siliciclastic shoaling, with a *Sphaerirhynchia wilsoni* Association succeeded by the specialised fauna of the *Shaleria ornatella* Association. The latter has a narrow stratigraphical range, but has been recorded throughout the Welsh Borderland. A temporary change of ocean current direction, resulting in a different pattern of larvae distribution has been suggested to account for its development (Watkins, 1979, p.236). Late Ludfordian beds contain a *Protochonetes ludloviensis* Association, its low diversity and coquinoid siltstone facies indicating a proximal setting with common storm events. At the base of the Přídolí Series is the Ludlow Bone Bed Member. This is interpreted as a lowstand regressive lag deposit, later reworked and deposited in scours and ripple troughs in the topmost surface of the Upper Ludlow Shale during a transgressive event. Rare examples of the late Ludfordian fauna occur in the Downton Castle Sandstone, but are confined to the beds immediately above the Ludlow Bone Bed. The Raglan Mudstone Formation represents deposition on an alluvial coastal plain subject to repeated emergence and occasional transgressive shallow-marine to brackish influxes.

The regional palaeogeographical context of Silurian depositional history is reviewed by Bassett et al. (1992).

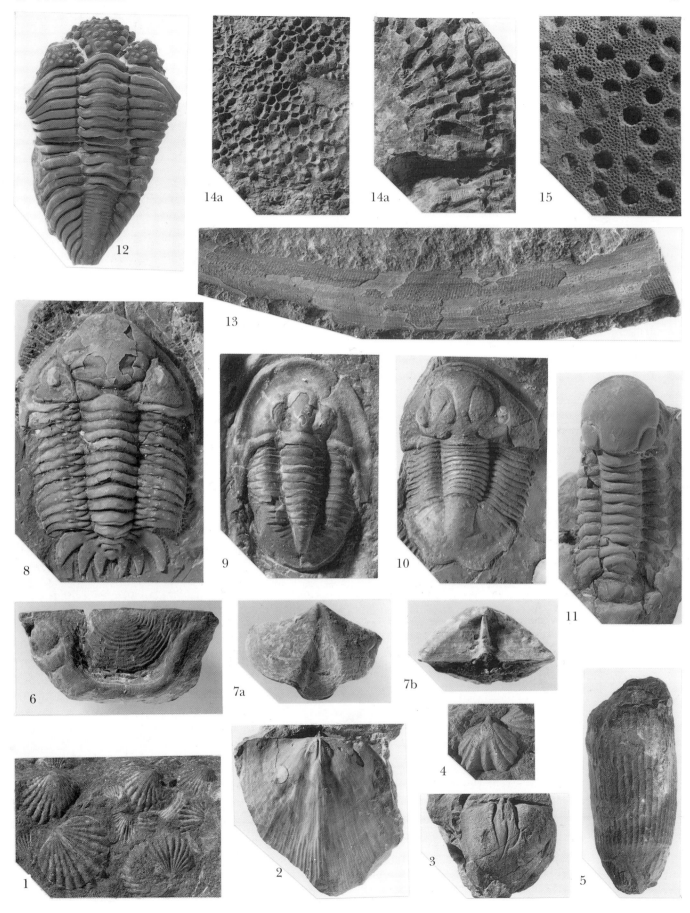

LLANDOVERY SERIES (MAY HILL SANDSTONE GROUP)

Llandovery Series rocks crop out in narrow belts to the west of the Malvern Hills, in the core of the Storridge Anticline around Old Storridge Common, and on Ankerdine Hill. They consist of a lower unit, the Cowleigh Park Formation, and the overlying Wyche Formation, which together make up the May Hill Sandstone Group (Sedgwick, 1853). The lower formation equates with the Huntley Hill Formation of the southern Malverns, the upper with the Yartleton Formation (Worssam et al., 1989).

First named by Groom (1910), the rocks were described in detail most recently by Ziegler, Cocks and McKerrow (1968). They represent a major transgressive episode of drowning of the Midland Platform; palaeoecological and sedimentological studies indicate deposition in progressively deeper water.

Evolutionary lineages of the brachiopods *Eocoelia* and stricklandiids have been described by Ziegler (1965) and Ziegler, Cocks and Bambach (1968). The stratigraphical relationships of these lineages to the graptolite biozones have since been revised by Cocks et al. (1984). In the Worcester district the stratigraphical range of the Llandovery rocks extends from the Aeronian Stage, based on records of *Eocoelia hemisphaerica* in the Cowleigh Park Formation, to the latest Telychian Stage, as indicated by the presence of *Eocoelia curtisi* and *Costistricklandia lirata* high in the Wyche Formation (Ziegler, Cocks and McKerrow, 1968). However, there is evidence of a non-sequence involving the lower part of the Telychian Stage at the junction of the two formations.

The junction of the Llandovery Series rocks and older rocks

The nature of this boundary has generated much discussion on whether it is a fault (Groom, 1899, 1900; Robertson, 1926; Raw, 1952; Whitworth, 1962; Phipps and Reeve, 1964, 1967, 1969) or unconformity (Phillips, 1848; Symonds, 1872, 1880; Reading and Poole, 1961, 1962; Butcher, 1962). Reading and Poole described excavations at Gullet Quarry (1.5 km south of the district) which exposed a basal Llandovery conglomerate resting on a weathered, uneven surface of the Malverns Complex. Phipps and Reeve (1964, 1967, 1969) and Whitworth (1962) argued that the exposures demonstrated that the contact was a fault, but Ziegler (1964), Shelford (1964), Tucker (1965), Brooks and Druce (1965), Jones et al. (1969), Brooks (1970), Hancock (in discussion of Brooks, 1970) and Brooks and Bassett (1970) supported Reading and Poole's interpretation of an unconformity, restated by McKerrow and Reading (in discussion of Phipps and Reeve, 1969). The opposing views were summarised by Penn and French (1971), who supported the unconformity interpretation.

The Gullet section is exposed today, and clearly shows an unconformity at this locality (Bullard, 1989; Worssam et al., 1989) There is no doubt, however, that at other localities the junction is a tectonic one. Reading and Poole (1962) concluded that 'we cannot say whether there is a continuous fault on the western side of the Malverns or not, but if there is, then at some localities it must lie either within the Malvernian or within the Llandovery; if within the latter then the throw must be less than 1000 feet. It is unlikely to be a considerable fault and certainly is not everywhere a *boundary* fault'. Brooks (1970) stated that there could be no reasonable doubt of the unconformable nature of the junction at the two localities then exposed, the Sycamore Tree exposure (Locality 7 below) and the Gullet Quarry, but concluded that the western boundary is 'undoubtedly a line of considerable tectonic disturbance'.

The transpressive model suggested by Chadwick and Smith (1988) implies that the junction is locally a high-angle thrust fault; the Llandovery rocks are overturned below the junction locally, supporting this interpretation. However, the occurrences of an unconformable junction imply that the vertical displacement on the thrust is not great, and that, as suggested by Reading and Poole, the thrust lies locally within the Llandovery rocks. The tectonic synthesis incorporating this interpretation is given in Chapter 10. The recorded and recent exposures of the junction within the Worcester district are listed below.

Plate 1 Selected Llandovery and Wenlock fossils (see also Plate 2) from the Malvern Hills.

All the specimens are in the Palaeontological Collections of the British Geological Survey.
Photographs by P Wells.

1. *Eocoelia hemisphaerica* (J de C Sowerby), GSd 4829 × 2. Cowleigh Park Formation, Ankerdine Hill.
2. *Costistricklandia lirata* (J de C Sowerby), GSM 13736, × 1. Wyche Formation, Howler's Heath.
3. *Mackerrovia lobatus* (Lamont and Gilbert), GSM 11461, × 1. May Hill Sandstone Group, Gunwick Mill, near Alfrick.
4. *Stegerhynchus decemplicatus* (J de C Sowerby), Geol. Soc. Coll. 6912, × 2. Cowleigh Park Formation, Ankerdine Hill.
5. *Kionoceras angulatum* (Wahlenberg), GSM 104540, × 1. Woolhope Limestone Formation, Wych Tunnel, Malvern.
6. *Leptaena sperion* Bassett, GSM 12681 (paratype), × 1. Woolhope Limestone Formation, Worcester Railway, near Malvern.
7a, b *Cyrtia exporrecta* (Wahlenberg), GSM 13588, × 2. Woolhope Limestone Formation, Malvern.
8. *Cheirurus centralis* Salter, GSM 36094, × 2. Coalbrookdale Formation, The Wych, Malvern.
9. *Prantlia grindrodi* Owens, GSM 3303 (holotype), × 4. Coalbrookdale Formation, The Wych, Malvern.
10. *Hemiarges scutalis* (Salter), GSM 19531, × 2. Coalbrookdale Formation, Malvern.
11. *Sphaerexochus mirus* Beyrich, GSM 36120, × 2. Coalbrookdale Formation, Malvern.
12. *Encrinurus tuberculatus* (Buckland), Z1 4103, × 2. Coalbrookdale Formation, Malvern.
13. *Ptilodictya lanceolata* (Goldfuss), GSM 87333, × 2. Much Wenlock Limestone Formation, Malvern.
14a, b *Favosites* sp., GSM 104538, × 2. Much Wenlock Limestone Formation, Malverns.
15. *Heliolites* sp., GSM 104539, × 2. Much Wenlock Formation, Malverns.

1. Newer railway tunnel, Colwall [7675 4325]; Robertson (1926, pp.165, 167) recorded a vertical fault juxtaposing vertical mudstones, 'broken, but not violently crushed for about 10 feet' against Malverns Complex hornblende schist. Despite Brooks' view (1970, p.252) that Robertson's account is 'rather uncritical', it seems likely that the junction is genuinely a fault. Robertson (1926, p.165) also described an inlier of Llandovery strata in the tunnel at 568 m (1865 feet) from the east portal, faulted between Malverns Complex rocks, and comprising 4.57 m (15 feet) of crushed calcareous mudstone. Robertson correlated this fault zone with the occurrence of Llandovery rocks recorded by Symonds and Lambert in the older railway tunnel to the north (Locality 2).

Plate 2   Selected Wenlock (see also Plate 1) and Ludlow fossils from the Malvern Hills.

All the specimens are in the Palaeontological Collections of the British Geological Survey. Those labelled 'Ledbury' may be from localities immediately south of the Worcester Sheet. Photographs by P Wells.

1. *Dalmanites 'caudatus'* (Brünnich), GSM 19325, × 2. Much Wenlock Limestone Formation, Malvern.
2. *Acaste downingiae* (Murchison), GSM 19352, × 2. Much Wenlock Limestone Formation, Dean Hill, Malvern.
3. *Staurocephalus susanae* Thomas, GSM 49805 (holotype), × 4. Much Wenlock Limestone Formation, Malvern railway tunnel.
4. *Conularia sowerbyi* de Varneuil, Geol. Soc. Coll. 6713, × 2. Much Wenlock Limestone Formation, Brand Lodge, Malvern.
5. *Dawsonoceras annulatum* (J Sowerby), GSM 104541, × ¹/₂. Much Wenlock Limestone Formation, Malvern.
6. *Grammysia* sp., GSM 24212, × 1. Much Wenlock Limestone Formation, Malvern.
7. *Dolerorthis rustica* (J de C Sowerby), GSM 13280, × 2. Much Wenlock Limestone Formation, Malvern.
8. *Meristina obtusa* (J Sowerby), GSM 12232, × 1. Much Wenlock Limestone Formation, Malvern.
9. *Orbiculoidea forbesii* (Davidson), GAM 16539, × 1¹/₂. Much Wenlock Limestone Formation, Winnals Farm, Malvern.
10. *Trigonirhynchia stricklandii* (J de C Sowerby), GSM 15960, × 1¹/₂. Much Wenlock Limestone Formation, Malvern.
11. *Slava fibrosa* (J de C Sowerby), GSM 24089, × 1. Lower Ludlow Shale, Ledbury.
12. *Proetus (P.) concinnus* (Dalman), GSM 36748, × 2. Lower Ludlow Shale, Cut-throat Lane, Ledbury.
13. *Dicoelosia biloba* (Linnaeus), GSM 13461, × 3. Lower Ludlow Shale, Cut-throat Lane, Ledbury.
14. *Glassia obovata* (J de C Sowerby), Geol. Soc. Coll., 6624 (lectotype), × 2. Lower Ludlow Shale, Malvern.
15. *Shagamella minor* (Salter), GSM 13559, × 3. Aymestry Limestone Formation, Malvern.
16. *Ptychopteria* sp., GSM 21830, × 1. Aymestry Limestone Formation, Ledbury.
17. *Salopina lunata* (J de C Sowerby), GSM 13381, × 2. Upper Ludlow Shale, Frith Farm, Malvern.
18, 19 *Microsphaeridiorhynchus nucula* (J de C Sowerby), GSM 16123, 16124, pedicle and brachial valves, respectively, × 2. Upper Ludlow Shale, Ledbury.
20. *Protochonetes ludloviensis* Muir-Wood, GSM 13578, × 2. Upper Ludlow Shale, Malvern.

2. Older railway tunnel, Colwall, 13.4 m (44 feet) north of the newer one; Symonds and Lambert (1861, pp.157, 158) recorded two sections. The first (their figure 4) is shown as a subhorizontal fissure in the Malverns Complex filled with 'two thin bands of Llandovery limestone, with strata of marly shales 2 feet thick'. Robertson (1926, p.165; see (1) and Phipps and Reeve (1964, p.406) referred to this as a fault. The main contact (Symonds and Lambert, 1861, fig.5) was described as comprising Upper Llandovery limestones and shales resting vertically against the Malverns Complex, with evidence of 'great pressure and crushing'.

3. Wyche cutting [c.7687 4372]; Phillips (1848, p.64) and Groom (1900, p.149) described a fault-bounded sliver of Llandovery rocks within the Malverns Complex.

4. By small roadside car park [7651 4557], West Malvern. A trench was dug across the outcrop of the junction in 1986, revealing a fault zone. However, some rounded cobbles and boulders of Malverns Complex rocks suggested the presence of a basal Llandovery conglomerate.

5. The Dingle (Phillips, 1848; Groom, 1900; Robertson, 1926, p.168); said to have exposed a vertical faulted junction, with the Llandovery rocks much crushed and schistose layers in the Malverns Complex dipping east-north-east at 70°.

6. Unpublished section of quarry exposures [c.7648 4568] recorded by Prof. A H Green (BGS field note-book, 1892) north of The Dingle shows Llandovery strata dipping steeply to the west and resting unconformably on the Malverns Complex (Figure 17). The overlying Llandovery strata were seen in the face of the

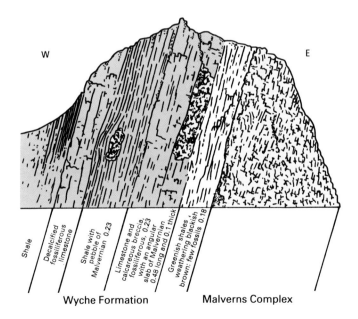

Figure 17   Sketch drawn by Prof. A H Green in 1892 of the unconformity between the Malverns Complex and the Wyche Formation. Locality lies north of the Dingle, West Malvern.

quarry and recorded by Green as 'an arkose of disintegrated sienite [*sic*], such as might have been formed on the spot, but with small, well rounded pebbles of white vein quartz'.

7. Small exposure [7647 4594], West Malvern; known in the literature as the Sycamore Tree exposure or Sycamore Tree Quarry, this is near the locality described (and figured) by Phillips (1848, pp.66, 67) and subsequently described by Robertson (1926, p.168), Reading and Poole (1961, 1962), Butcher (1962), Shelford (1964, pp.566, 567), Hardie (1969, p.11) and Penn and French (1971, p.33). It is now much overgrown, but a poor exposure was visible in 1982. Like the Gullet Quarry and Green's section to the south (Locality 6), this shows an unconformity between the Malverns Complex and the basal Llandovery beds (e.g. Brooks, 1970). The latter rest on an irregular surface of diorite, and comprise two conglomerates separated by up to 0.5 m of grey shale. The lower conglomerate contains Malverns Complex boulders, the upper is richly fossiliferous. The conglomerate is known as 'Miss Phillips Conglomerate' after Phillips' sister who discovered it (Phillips, 1848, pp.66, 67). A unique assemblage of conodonts is recorded from the conglomerate (p.26). The surface of the Malverns Complex seen in 1982 dips steeply to the west; Phillips (p.67) described the surface as nearly vertical, but undulating and irregular. Shelford (1964, p.567) and Hardie (1974) noted a westerly dip, Penn and French (1971) an easterly one.

8. Broomhill Quarry [7641 4656]; described and figured by Groom (1900, pp.151, 152 and fig.7) as the old quarry north of St Edward's Orphanage. He showed a fault complex comprising fault-bounded slices of Malverns Complex rocks and Llandovery rocks. A trench was excavated by Bullard (1989, pp.36, 37) in 1986. This confirmed Groom's record, exposing a 4 m zone of highly disturbed Llandovery beds between Malverns Complex rocks.

9. Sketch section (unpublished BGS notebook) made by Prof. A H Green in 1892 records temporary exposures in an excavation [c.7642 4660] for a tennis court at Broomhill, West Malvern. This shows a steeply dipping, inverted sequence of interbedded unconsolidated purple grits and red mudstones resting on crystalline rubble, 'apparently the weathered surface of solid rock'. A layer of dark tough clay separates the sedimentary rocks from the crystalline ones, but it is not clear whether this is of sedimentary or tectonic origin. Green refers to the bed as '?leathercoat', suggesting that it is sheared gouge, but his description of the section would indicate that the junction is basically an unconformity. The grits are described by Green as comprising 'angular and subangular fragments of crystallines, averaging a small pea in size, in gritty matrix, which is probably only crystallines broken smaller'. The basal grit contains larger, angular fragments. The sequence is typical of the red-bed facies of the Cowleigh Park Formation.

10. Former quarry [?c.7654 4710] described by Groom (1900, pp.152, 153) as situated 'close to the eastern side of the upper road, at a point about due north-east of the Lamb Inn' [7646 4701]; Groom (Figures 8 and 9) interpreted exposures as belonging to a fault zone comprising three parallel faults, two of which are thrusts, with a fault-bounded inlier of late Llandovery, greenish, micaceous sandstones to the east of the main western fault of the Malverns Complex. The fault strikes north 3° west (not east, as stated by Bullard, 1974), and dips about 60° to the east.

11. Temporary section [7656 4717], West Malvern, seen in 1982, showed thinly bedded Llandovery Series mudstones and sandstones overturned and dipping about 70° to the east-north-east, faulted against Malverns Complex diorite which is sheared to a soft mica schist at the fault. The fault strikes at N30°E and dips about 80° to the east.

12. Temporary section [7655 4730] recorded by Bullard (1974) exposed red beds of the Cowleigh Park Formation, faulted against Malverns Complex diorites and dolerites; the junction is marked by a thin layer of black shale dipping 65° to the east (p.15). Bullard concluded that the section disproved the theory of a major western boundary fault to the Malverns Complex, as proposed by Phipps and Reeve (1964, 1969) and supported the theory of pre-Llandovery tectonics postulated by Brooks (1970), thereby implying that the junction here is an unconformity. However, his figures 2 and 3 illustrate a fault complex similar to those described at localities 10 and 11.

13. Temporary sections and boreholes [7670 4745] at Cowleigh Bank examined in 1979 by R A Ellison and B S P Moorlock, following a slope failure during excavations for a service reservoir (p.94). Both unconformable and faulted contacts were observed; red pebbly mudstones of the Cowleigh Park Formation unconformably overlie Malvern Complex rocks locally, green mudstones are faulted against the Malverns Complex locally, and in one case (p.15), Cambrian black shale overlies the Malverns Complex.

14. Temporary section [7662 4751] Cowleigh Park; sheared Malverns Complex granite was seen to be faulted against purple-red friable granule- and pebble-rich mudstones of the Cowleigh Park Formation, with a blue-grey clay along the fault, which strikes at about 100°.

15. Road cutting [7632 4663], Cowleigh Park; Phillips (1848, p.37) illustrated a section, which was patchily re-exposed during road widening in 1987 (Figure 14). The cutting lies in a fault zone, and although Phillips showed only the western boundary of the Malverns Complex to be faulted, the eastern boundary, juxtaposing tough pale grey quartzite correlated with the Cambrian Malvern Quartzite (p.15), is also likely to be faulted. Groom (1900, p.156) also referred to this section.

The only other known occurrences of the junction between the Llandovery rocks and those of the Cambrian are described on p.15.

Cowleigh Park Formation

The Cowleigh Park Formation is restricted to a narrow outcrop extending northwards from West Malvern (where it is overstepped by the Wyche Formation) to its type area of Cowleigh Park, a tract in the core of the Storridge Anticline in the Birchwood–Old Storridge Common area, and to Ankerdine Hill.

The formation is mainly arenaceous, consisting of purple, brown, buff and green sandstones and conglomerates with minor mudstones. In the type area, purple sandstones and purple hematitic granule-rich and pebbly mudstones and sandstones predominate.

Previous estimates of thickness of the formation are 107m to 122 m (350 to 400 feet, Ziegler, Cocks and McKerrow, 1968), over 152 m (500 feet, Phipps and Reeve, 1967), 213 m (700 feet, Phillips, 1848), and 274 m (900 feet, Groom, 1910). Although faulting complicates the situation, the outcrop width from High Wood to North Malvern is 550 m, and dips range from 32° in High Wood to near vertical at North Malvern, giving a minimum thickness of 300 m and a maximum of perhaps 450 m.

Outcrops of the upper beds on Old Storridge Common and Ankerdine Hill contain *E. hemisphaerica* (Ziegler, Cocks and Bambach, 1968), indicating a late Aeronian age. A similar age is likely in the Malvern area, although no stratigraphically useful fossils have been recorded.

There are few good sections, although small temporary exposures have been recorded. The best is in a bank [7617 4724] by the football pitch in High Wood, West Malvern (Plate 3). Described by Ziegler, Cocks and McKerrow (1968, p.753), the section was cleared in 1986 (Bullard, 1989), but has now deteriorated again. Nevertheless, the main elements can be recognised. The beds lie at the top of the formation, and apparently pass upwards into the basal beds of the Wyche Formation (p.27). A sequence of lenticular, hematite-rich, red beds comprise harder, cemented, fluvial, channel-fill sandstones and softer, weakly cemented or uncemented, granular breccia layers. The sandstones are conglomeratic, with an abundance of small hematite-coated angular to subangular sandstone clasts in a gritty purple-red matrix. The granular layers are dark purple to metallic blue-grey beds of patinated granules and small angular to subangular fragments of Malverns Complex granite and sandstone, all coated with hematite. They resemble unconfined flashflood fluvial deposits, which, together with the channelised sandstones, were probably deposited in proximal fans emanating from the Midland Platform nearby, to the east. Ziegler, Cocks and McKerrow (1968) considered these unusual beds to be probably confined to the Cowleigh Park area, adjacent to the Malvernian source, and to perhaps have been beach deposits. However, the angularity of the clasts, the unsorted nature of the breccias and the channelised sandstones all point to a fluvial origin. Similar red beds occur locally at the base of the formation. The higher occurrences indicate a regressive event and provide sedimentological evidence to support the faunal evidence of Ziegler, Cocks and McKerrow (1968) for a non-sequence at this level.

Close to the east of the football pitch section, quartz-rich sandstones and conglomerates form a prominent ridge which extends from here north to Cowleigh Park. This is displaced dextrally, about 200 m, by the Cowleigh Park fault system and continues northwards over Rough Hill Wood. The best section [7600 4792] is by the Cowleigh Road, where 4 m of greenish grey, fine-grained, quartzitic sandstones and quartz grits with thin brown mudstone interbeds are exposed. Phipps and Reeve (1969, plate 1) correlated these beds with 'Miss Phillips' Conglomerate', but they appear to lie at a stratigraphically lower level than the basal Llandovery conglomerate at Sycamore Tree Quarry [7646 4594], West Malvern (p.23), which is the type locality for 'Miss Phillips' Conglomerate'.

The formation crops out along the steep, mainly wooded slopes of Old Storridge Common [748 511], where about 110 m are present. The outcrop is strewn with sandstone debris, but exposures are mainly confined to the Leigh Brook and its tributaries. A partially filled, disused quarry [7478 5118] in purple and green, coarse-grained, feldspathic sandstone contains *Lingula* sp., *Eocoelia hemisphaerica* and bivalves (Lamont and Gilbert, 1945; Ziegler, Cocks and McKerrow, 1968).

The topmost beds of the formation crop out in the core of the Malvern Axis on Ankerdine Hill and consist of hard green sandstones with some mudstones. There are only a few small exposures.

Wyche Formation

The Wyche Formation crops out in a narrow belt of steeply dipping rocks close to the west side of the Malverns, from High Wood north to Old Storridge Common, and on the flanks of Ankerdine Hill.

The formation consists predominantly of grey and pale green mudstones and siltstones, but contains numerous thin, tabular, green, fine-grained sandstones. The mudstones and siltstones weather to sticky grey and buff clays at the surface. Bentonites are common. The sandstones are closely jointed, and the outcrop is strewn with small tabular sandstone fragments. The weakness of the clays at the surface results in shallow earth flow land-slipping in many parts of the outcrop. The formation includes, at its top, the beds referred to in the earlier literature as the Woolhope Shale (Groom, 1900; Phipps and Reeve, 1967).

A maximum of 150 m of strata of the Wyche Formation are estimated to be present in the district.

The basal conglomerates at the Sycamore Tree Quarry, West Malvern and those at the base of the Wyche Formation at Gullet Quarry contain many species absent elsewhere in the Welsh Borderland (Ziegler, Cocks and McKerrow, 1968), interpreted as a mixture of rocky bottom and soft bottom species. In the Old Storridge Common and Ankerdine Hill areas the basal beds of the formation contain a *Pentameroides* Community, which, with *Eocoelia curtisi* recorded at the latter (Ziegler, Cocks and McKerrow, 1968), indicate a late Telychian age. With the topmost beds of the underlying Cowleigh Park Formation yielding *E. hemisphaerica*, the palaeontological evidence suggests a non-sequence at this level, with the lower part of the Telychian Stage not represented. The terrigenous beds at the top of the Cowleigh Park Formation indicate a regressive event, but whether the hiatus is an erosional one is not clear. The relationship between the formations may be one of onlap, with the Cowleigh Park Formation filling an irregular Malvernian topography, the latter being overstepped completely from West Malvern southwards, where the Wyche Formation rests unconformably on, or is in faulted contact with, the Malverns Complex. However, it is also possible that the two formations have a more complex relationship, with terriginous red beds of the Cowleigh Park Formation interfingering with the marine Wyche Formation.

Beds high in the formation near Mousehole Bridge [7400 5153] contain *Eocoelia sulcata* of latest Telychian age, and in the Malvern area, a *Pentameroides* Community in the basal beds is succeeded by beds with *Costistricklandia lirata*, again indicating a latest Telychian age (Ziegler, Cocks and McKerrow, 1968). A *Clorinda* Community fauna was recorded by the same authors in the upper part of the formation at West Malvern.

Two occurrences of the *Petalocrinus* Limestone have been recorded (Pocock, 1930). This lies in the topmost part of the formation in beds formerly ascribed to the Woolhope Shale, and compares closely in thickness and stratigraphical position, just below the Llandovery–Wen-lock Series boundary, to its occurrence at Woolhope, where it is about 0.03 m thick and rests on a 0.1 m-thick limestone with large tabulate corals.

Favosites gothlandicus, Atrypa sp., *Leptaena rhomboidalis* and *Resserella* [*Parmothis*] cf. *elegantula* were recorded by Phipps and Reeve (1967), who noted that plate-like colonies of *F. gothlandicus* are a characteristic feature of the top of the formation.

In the conodont assemblage recorded by Aldridge (1972) from the conglomeratic base of the formation at Sycamore Tree Quarry, the dominant species is *Icriodella malvernensis*, which has only been recorded from this locality. The associated fauna includes *Ozarkodina gulletensis, O. aldridgei, Icriodella inconstans, Kockelella ranuliformis* and *Distomodus staurognathoides* (Aldridge, 1972, tables I–IV), suggesting assignment to the *D. staurognathoides* conodont Biozone and a late Aeronian or low Telychian age. A conodont fauna, similar to that collected from Sycamore Tree Quarry, but lacking *I. malvernensis* and *O. aldridgei*, has been recovered from low in the formation in a tributary to Leigh Brook [7438 5114], south of Alfrick (Locality 2; R J Aldridge, unpublished collections).

Conodonts of the *Pterospathodus amorphognathoides* Biozone were reported (Aldridge, 1972) from the uppermost beds of the formation at Birches Farm Lane [7603 4685], West Malvern (Locality 10). A detailed study of microfossil assemblages was carried out by Mabillard (1981), who noted that *P. amorphognathoides* ranges from the upper Wyche Formation into the basal 1.7 m of the

Plate 3 Beds of the Cowleigh Park Formation, Cowleigh Park; thinly bedded sandstones and granule beds. Photograph by Dr J S W Penn.

Woolhope Limestone. In the type section for the base of the Wenlock Series at Leasows, Shropshire, the highest record of this species is 30 cm above the Llandovery–Wenlock series boundary (Mabillard and Aldridge, 1985). A correlation of the Birches Farm section with the Llandovery–Wenlock boundary beds is supported by the occurrence of the acritarchs *Domasia amphora* and *Deunffia brevispinosa* in the lower part of the Woolhope Limestone (Mabillard, 1981). These species are diagnostic of acritarch zone 5 (Hill, 1974; *Deunffia brevispinosa* and *Deunffia brevifurcata* biozones of Dorning and Bell, 1987), the base of which occurs 15 cm below the base of the Wenlock Series at Leasows (Mabillard and Aldridge, 1985). The ostracods *Menoeidina lavoiei* and *Hemiaechminoides monospinus* occur in the basal metre of the Woolhope Limestone at Birches Farm, but are found only in strata of Llandovery age at Leasows (Mabillard, 1981). The combined microfossil evidence suggests that the base of the Wenlock Series is about 1 m above the first nodular limestones of the Woolhope Limestone at Birches Farm.

There are few good sections of the Wyche Formation, but numerous small exposures. The following lists the better exposures.

1. Exposures [7407 5264; 7434 5269; 7440 5264] of beds low in the formation at Workhouse Bank. A palynological sample (MPA 25226) from the first exposure yielded a sparse to moderate assemblage of acritarchs dominated by *Diexallophasis denticulata* and *Cymatiosphaera octoplana*. Stratigraphically useful taxa include *Dilatisphaera dameryensis*, *Domasia trispinosa*, *Salopidium granuliferum* and *Tylotopalla robustispinosa* which indicate a mid-Aeronian to Telychian age (*Oppilatala eoplanktonica* to *Deunffia monospinosa* biozones of Dorning and Bell, 1987). The second locality (MPA 25237) yielded a sparse to moderately abundant assemblage of acritarchs including *Ammonidium microcladum*, *Cymatiosphaera llandoveryensis*, *Diexallophasis pachymura*, *S. granuliferum* and *Tylotopalla digitifera*. These suggest a similar mid-Aeronian to Telychian age (*Oppilatala eoplanktonica* to *Deunffia monospinosa* biozones of Dorning and Bell, 1987). The third locality (MPA 25236) yielded a diverse and abundant assemblage dominated by *D. denticulata* and *D. trispinosa*. The presence of *Domasia bispinosa*, *Domasia limaciformis* and *Gracilisphaeridium encantador* indicate a Telychian to earliest Sheinwoodian age (*Deunffia monospinosa* to *Deunffia brevifurcata* biozones of Dorning and Bell, 1987). Evidence from central Wales suggests that *G. encantador* first appears within the upper part of the *turriculatus* graptolite biozone (Barron *in* Fletcher et al., in press).

The first exposure is of 1.5 m of mudstones, weathered to sticky clay, interbedded with hard siltstones and very fine-grained sandstones in beds up to 0.08 m thick. The second yielded the brachiopods *Craniops implicatus*, *Dicoelosia* sp., ?*Leangella* and ?*Sphaerirhynchia*, the third *Coolinia applanata* and *Jonesea grayi* and the trilobite *Encrinurus diabolus*?.

2. Beds at the base of the formation are exposed in Leigh Brook [7430 5152] and a tributary [7440 5121] where they rest on the underlying Cowleigh Park Formation (p.25). They consist of interbedded mudstones, silt-

stones and sandstones; the mudstones are weathered to ochreous clay. The sandstones are pale grey-green and typically less than 0.04 m thick, but one is 0.1 m thick and contains abundant brachiopods.

3. Beds high in the formation are exposed at two localities [7405 5167; 7402 5153]. They consist of purplish and yellow-green silty mudstones and siltstones with a few thin very fine-grained sandstones. Ziegler, Cocks and McKerrow (1968) recorded a *Clorinda* Community fauna from the first locality.

4. Scattered stream exposures [7432 5101–7440 5120] in Coneygore Coppice, mainly of mudstones; at one locality [7438 5114] Ziegler, Cocks and McKerrow (1968) recorded a *Costistricklandia* Community fauna.

5. A waterfall exposure [7452 4998] in Woodhouse Dingle of 0.2 m of grey-green, fine-grained, tough, tabular sandstone above grey to blue-grey mudstones with thin sandstones. Mudstones with thin sandstones are also exposed above the waterfall.

6. Storridge road cutting [7570 4907–7546 4887], where the beds crop out in the core of the Storridge Anticline. The cutting is now largely grassed over, but there are a few exposures of mudstones and interbedded thin fine-grained sandstones, as well as one of a red and white mottled bentonite.

7. Stream exposures [7538 4861–7539 4854] mainly of fine-grained, parallel-laminated sandstones.

8. Stream bank exposures [7581 4786] in Whippets Brook of beds at the top of the formation; 5 m of purplish brown mudstones with some thin sandstones and a layer of lilac bentonite are exposed on the north bank; on the south bank beds immediately underlying the Woolhope Limestone consist of green silty mudstones with a few limestone layers and nodules. Small exposures of sandstones, limestones and mudstones are present to the east. The limestones contain *Costistricklandia* sp. and *Coolinia* sp.

9. Section [7616 4724] in bank by football pitch, West Malvern, exposing the basal beds of the formation and their junction with the Cowleigh Park Formation (p.25; Ziegler, Cocks and McKerrow, 1968). About 15 m of pale grey-green and buff clayey mudstones and tabular sandstones are exposed. The sandstones are mainly green and fine grained, but the basal ones are purple and red-brown and gritty like those in the underlying Cowleigh Park Formation. They are typically about 0.05 m thick, but some are up to 0.1 m and one is 0.8 m. Bullard (1989) recorded two thin fossiliferous limestone layers and a pale grey bentonite when the section was cleared in 1986.

10. Trackside section [7603 4685] at Birches Farm, West Malvern, exposes the topmost 1m of the formation, comprising silty mudstone with a 0.08 m fine-grained sandstone near the top, and the overlying Woolhope Limestone. Microfossils recovered from the section are listed on p.26.

11. Temporary sections [7624 4632; 7625 4626] at building sites in West Malvern exposed interbedded mudstones and sandstones. The mudstones are green, weath-

ering yellow, with some dull purplish brown and red-brown layers. The sandstones are about 0.05 m average thickness, ranging up to 0.1 m. They have sharp tops and bases and contain abundant worm traces.

12. In the Colwall railway tunnel, Robertson (1926) recorded about 76 m (250 feet) of flaggy, calcareous and sandy, grey to purple and mottled brown mudstones. Sandstones are uncommon in any part of the sequence, but rare near the top. Shelly, brachiopod-rich bands are common, and *Pentamerus oblongus* and *Stricklandia lens* were recorded.

WENLOCK SERIES

Rocks of the Wenlock Series crop out in a narrow zone west of the Malverns and in the core and flanks of the Malvern Axis to the north. They comprise three formations, the Woolhope Limestone Formation, the Coalbrookdale Formation and the Much Wenlock Limestone Formation.

Woolhope Limestone Formation

The Woolhope Limestone consists of thinly bedded, nodular limestones and calcareous siltstones and mudstones. The limestones range from crinoidal bioclastic grainstones to argillaceous wackestones, the latter being most common. They are blue-grey where fresh, green-grey to buff where weathered. The interbedded siliciclastic layers are buff to olive-green. Thin bentonite layers also occur. Over much of its outcrop the formation makes a weak landform feature. Equivalent to the Buildwas Formation of the Wenlock Edge area, it represents the earliest part of the Sheinwoodian Stage.

The formation ranges from about 6 m to 18 m in thickness, although over much of its outcrop it maintains a relatively uniform thickness of about 15 m, perhaps thinning regionally northwards from 17 m in the Colwall tunnel (Robertson, 1926). The minimum thickness of 6 m was recorded in a disused quarry [7490 4914] at Merryhill Common, Storridge. No evidence was found for the presence of two calcareous members separated by mudstones, nor of these mudstones attaining a thickness of 45 m (150 feet) in the North Malvern area, as proposed by Phipps and Reeve (1967, p.343). Instead, it appears that, as in the Collington Borehole [6460 6100] and in the Eastnor Park Borehole [7437 3809] (Worssam et al., 1989), the formation consists of several clean limestones within an argillaceous sequence. The section in a shallow cutting on the A4103, Storridge, comprising about 8 m of interbedded calcareous siltstones and limestones, is probably typical of the formation throughout the district.

Crinoid fragments and shell debris are common constituents. The macrofauna tends to be sparse and lacks diversity. Small brachiopods, together with tabulate and rugose corals, including *Schlotheimophyllum patellatum* are most commonly recorded, molluscs and trilobites being poorly represented. The brachiopods are mainly small

forms and include *Atrypa reticularis*, *Cyrtia exporrecta*, *Dalejina hybrida*, *Dicoelosia biloba*, *Eoplectodonta duvalii* and *Leptaena* sp. (Phipps and Reeve, 1967). Hurst (1975a) recorded *Resserella sabrinae sabrinae*. Bassett (1974, p.760) reported *Eocoelia* cf. *sulcata*, though this is likely to be *E. angelini*, which is characteristic of the early Sheinwoodian.

The rich conodont fauna of the uppermost Wyche Formation diminishes abruptly in the lower part of the Woolhope Limestone, and 2 m above the base only coniform taxa are common (Mabillard, 1981). No age-diagnostic conodonts have been recovered from the middle or upper part of the formation in the north Malverns.

The comminuted nature of much of the bioclastic remains suggested to Phipps and Reeve (1967) that deposition took place in shallow water above wave base. The mixed carbonate–siliciclastic nature of the deposits may, however, indicate a more distal setting, with storms carrying material out from the shallower carbonate-producing areas.

The better exposures of the formation are: roadside exposures [7397 5292] west of Workhouse Bank, Suckley (Penn and French, 1971); on the western limb of the Storridge Anticline, Suckley [7396 5248; 7409 5185; 7388 5153; 7421 5079; 7424 5053–7432 5006; 7435 5010]; lane-side section at Alfrick Pound [7447 5192]; roadside exposures [7451 5216–7472 5214] east of Alfrick (Penn and French, 1971); disused quarry and adjacent cutting [7490 4914], Merryhill Common; exposures in track [7562 4925–7568 4933] on the eastern limb of the Storridge Anticline, Storridge; exposures in track [7509 4874–7515 4875] on the western limb of the Storridge Anticline, Storridge; exposures [7525 4860] in shallow cutting on the A4103, Storridge; disused roadside quarry [7542 4845]; good bedding plane exposures [7562 4817], Hill Farm, Storridge; stream exposures [7578 4786]; trackside section [7602 4685], Birches Farm, West Malvern; new exposure [7642 4253] at Linden; disused quarry [7645 4116] south of Hawkelts Coppice.

In the Colwall railway tunnel, Robertson (1926) placed the top of the formation at a clay layer overlying crinoidal limestone, above which the beds become more argillaceous.

Two bentonites within 2 m of siltstones are exposed near Suckley [7427 5034], and two thin examples are present in the section at Linden [7642 4253].

Coalbrookdale Formation

The Coalbrookdale Formation, formerly known as the Wenlock Shale, consists of olive-green mudstones, silty mudstones and siltstones. Limestone nodules and layers occur throughout, but are commonest in the transitional beds at the base and top of the formation. Thin bentonites occur throughout; Penn (1969) recorded fifty-five in a temporary section at Alfrick. About 225 to 250 m of beds are present, generally occupying the low ground between the ridges of the Woolhope Limestone and Much Wenlock Limestone formations.

No stratigraphically useful fossils have been recorded from the formation within the district, but in its type area in Shropshire it has a stratigraphical range from mid-Sheinwoodian to late Homerian.

Although reported to be poorly fossiliferous by Phipps and Reeve (1967), a diverse shelly fauna, mainly of brachiopods, corals and trilobites, has been listed by these authors, and by Robertson (1926) and Penn (1969). Brachiopods include the small forms *D. hybrida*, *Dicoelosia biloba* and *Eoplectodonta duvalii*, as well as strophomenids such as *Amphistrophia funiculata*, *Leptaena* spp. and *Strophonella euglypha*. Corals recorded include tabulate and rugose forms, such as large examples of *Ketophyllum* sp., and trilobites include species of *Acaste*, *Calymene* and *Proetus*.

The following lists the main exposures:

1. Green mudstones and silty mudstones occur in a thrust wedge north of Collins Green [743 578]. They are overlain by Much Wenlock Limestone (p.30), and are thrust over Llandovery Wyche Formation beds. If the junction with the limestone is a normal stratigraphic one, these beds belong to the Coalbrookdale Formation. However, the fauna collected is not diagnostic, and could equally belong to the Lower Ludlow Shale, in which case the junction would be a thrust. Exposures in a disused pit [7434 5780] of green calcareous silty mudstone and siltstone with limestone layers yielded the trilobite *Dalmanites*, the brachiopods *Atrypa reticularis*, *Eospirifer* sp., *Gypidula lata*, *Protochonetes minimus*, *Strophochonetes* sp. and *Strophonella?*, and the bivalve *Orthonota rigida*.

2. Roadside exposures [7442 5268–7448 5276] near Alfrick of olive-green calcareous siltstone and silty mudstone in which *Dalmanites* is common. An indeterminate graptolite was also collected. Penn and French (1971, p.35) recorded the coral *Ketophyllum* and the small brachiopods *Antirhynchonella* and *Dicoelosia*.

3. Exposures along the stream [7397 5211–7386 5168] west of Barley Farm, and along Leigh Brook [7359 5146–7352 5123] of olive-grey calcareous fossiliferous silty mudstones and siltstones with a few thin beds and nodules of limestone and bentonite layers.

4. Roadside exposures [7404 5083–7378 5073] and stream exposures [7378 5068–7428 5004] of beds similar to those of (3).

5. Roadside exposures [7473 5215] near Alfrick Pound of beds near the top of the formation of green-grey calcareous siltstones, from which sporadic small brachiopods and a trilobite pygidium have been recorded.

6. Stream exposure [7542 4963] of 1.5 m of green mudstones with some limestone nodules.

7. Stream exposures [7460 4942; 7461 4939] of green-grey fossiliferous mudstone with limestone nodules.

8. Exposures [7466 4929] of green mudstones in old track.

9. Exposures [7480 4900] in road banks of green-grey mudstones with some limestone nodules.

10. Exposures [7450 4846] in Storridge road cutting of grey-green silty mudstones and siltstones with limestone layers at the top of the formation. Bullard (1989, p.54)

recorded several bentonite layers and the brachiopods *Atrypa*, *Dicoelosia*, *Leptaena* and *Resserella* at this locality.

11. Whitman's Hill Quarry [7485 4830], where the top-most 12 m of beds are exposed (p.30); they consist of calcareous siltstones with two thin bentonites.

12. Stream exposures [7576 4713–7576 4708] of tectonically disturbed green mudstones and silty mudstones with limestone nodules.

13. Temporary exposures [7639 4780] in pit for new reservoir dug in 1985, Cowleigh Park; numerous bore-holes also proved the beds; they consist of green calcareous silty mudstones and hard siltstones with a few limestone layers. The beds are generally poorly fossiliferous.

14. Shallow excavations [7648 4457; 7650 4451] at Park Wood in overturned buff siltstones.

15. Small exposures [7665 4345] in the track from Upper Wyche to Colwall Stone of green calcareous silty mudstones and siltstones.

Symonds and Lambert (1861) noted that the beds were exposed for 688 m in the Colwall rail tunnel and its western portal cutting, but these included Much Wenlock Limestone and Lower Ludlow Shale beds miscorrelated by them as Coalbrookdale Formation. In the second tunnel, Robertson (1926) recorded the beds, present for 712 m, as greyish blue mudstones, locally with a purple tint, and with a few calcareous and fossiliferous layers. Because of folding and faulting he found it difficult to estimate the thickness of beds present, but suggested that the lower 150 m might be present, with about 100 m of the overlying beds cut out by faulting.

Much Wenlock Limestone Formation

The Much Wenlock Limestone Formation (formerly the Wenlock Limestone) is easily recognised by the bold ridge it makes at outcrop. There are numerous sections and exposures in many disused pits and quarries.

Previous estimates of thickness of the formation include 30 to 50 m (100 to 180 feet, Phillips, 1848, p.78) and 76 to 91 m (250 to 300 feet, Groom, 1900, p.171). About 90 to 100 m are present on the western limb of the Malvern Axis and about 50m on the eastern limb at Cowleigh Park. Outliers on the eastern limb of the axis are also present at Limekiln Coppice [757 496] near Storridge and Collins Green [740 573].

The formation consists predominantly of flaggy, nodular, thinly bedded, pale blue-grey bioclastic limestones with thin calcareous mudstone interbeds. The limestones are mainly grainstones and wackestones, with some packstones, that have undergone stratified cementation and selective diagenetic compaction (Bathurst, 1987). Some pisolitic limestone occurs near the base of the formation; coquinal layers and burrowing occur throughout. Phipps and Reeve (1967) noted a sandy limestone facies near the top of the formation in Mathon Park [761 453]. Three small patch reefs have been recorded near the base of the formation (Phipps and Reeve, 1967; Penn, 1969, 1971; Scoffin, 1971). They are mainly micritic and similar in structure, textural composition and organic framework to those at Wenlock Edge. Tabulate corals, especially species of *Favosites*, *Halysites* and *Heliolites*, stromatoporoids and bryozoans were the principal reef builders, their skeletal parts being bound by encrusting tabulate corals such as *Alveolites* and *Thecia*, stromatoporoids and algae including stromatolites, *Girvanella* and *Rothpletzella*. The petrography of limestones from Upper Hall Farm quarry [716 382], near Ledbury, 1.5 km south of the district boundary, has been described by Bathurst (1987).

Deposition occurred on a shallow shelf, perhaps with surface irregularities of low amplitude (Phipps and Reeve, 1967). The coquinal limestone, pisolitic limestone and reefs were formed in shallow agitated water, and the more massive of the nodular limestones also exhibit evidence of wave and current action.

An abundant, diverse, shelly macrofauna has been recorded by Phipps and Reeve (1967). It includes corals, stromatoporoids, bryozoans, brachiopods and trilobites, but molluscs are generally poorly represented. Of the brachiopods, Phipps and Reeve list the following species as abundant or common: *A. funiculata*, *Atrypa reticularis*, *Coolinia pecten*, *D, hybrida*, *Gypidula* spp., *Howellella elegans*, *H. subinsignis* [as *Delthyris elevata*], *Leptaena* sp., *Leptostrophia filosa*, *Meristina obtusa*, *Microsphaeridiorhynchus nucula*, *Plectatrypa imbricata*, *Resserella canalis* [as *R. elegantula*], *Rhynchotreta cuneata*, *Sphaerirhynchia wilsoni*, *Stegerhynchus borealis* and *Strophonella euglypha*.

An unpublished conodont fauna from a sample collected by Dr A T Thomas from a disused quarry [7582 4599] 200 m south-west of Croft Farm, West Malvern, includes *Ozarkodina bohemica bohemica*, *O. excavata*, *Panderodus equicostatus* and *Pseudooneotodus* sp. This association represents the *O. bohemica bohemica* Biozone, which spans the upper Sheinwoodian to lowermost Gorstian stages (Aldridge and Schönlaub, 1989).

The following lists the best of the numerous sections in the district:

1. Disused quarry [7402 5737], Collins Green; up to 5 m of Permo-Triassic conglomerates (p.48) rest unconformably on nodular to rubbly bedded, fine-grained limestones. The limestones are pale green, but much reddened below the unconformity; they are argillaceous towards the top, and a cream bentonite up to 0.2 m thick occurs about 3 m from the top. About 12 m of generally poorly fossiliferous limestones are present, in which *Favosites?*, *Leptostrophia*, rhynchonelloids, bryozoans and crinoid fragments were noted. Vuggy fine-grained limestones are the lowest ones seen.

2. Disused quarry [7369 5066] behind an engineering works, Longley Green; much of the face is unsafe for detailed examination and machinery hampers access. About 55 m of beds are exposed, comprising thinly bedded to nodular limestones with siltstone and silty mudstone interbeds and a few bentonites. Fossils include the corals *Favosites* and *Heliolites*, the brachiopods *Atrypa reticularis*, *Gypidula* sp., *Isorthis* sp., *Leptaena depressa*, *Lingula* sp., *Meristina obtusa*, *Eoplectodonta* sp., and *Sphaerirhynchia wilsoni*, a few trilobites and crinoid fragments.

3. Whitman's Hill Quarry [7484 4830] exposes about 25 m of the basal beds of the formation as well as the topmost beds of the Coalbrookdale Formation (p.29). The limestones are thinly bedded to nodular and fine upwards from coarse crinoidal grainstones at the base to fine- to medium-grained grainstones/packstones. More argillaceous beds in the middle of the exposed sequence form a conspicuous hollow. Penn (1969, 1971) described small patch reefs near the base of the formation consisting of grey micrite with corals, bryozoa, stromatoporoids, brachiopods and algae. He also recorded pisolitic limestone near the base.

4. Vinesend Quarry [7517 4760] exposes fossiliferous rubbly to nodular bedded grainstones. The best section, in the lower part of the quarry, exposes 5 m of limestones in three posts; the lowest is 3 m thick and the upper two are each 1 m thick. Pale green-grey shale interbeds occur throughout. Bullard (1989, pp.55, 56) recorded abundant brachiopods including *Atrypa*, *Leptaena*, *Resserella* and *Strophonella*, the corals *Favosites* and *Syringopora*, fragments of the trilobites *Calymene* and *Dalmanites*, crinoid debris at some levels and trace fossils in the upper shaly partings.

5. Disused quarry [7640 4428], Park Wood: thinly bedded, nodular and tabular limestones are exposed, as well as pisolitic limestones, in a 200 m-long pit; Penn (1969, 1971) described patch reefs here consisting of coalescent lenses of shelly micrite in which growing organisms occupy about 25 per cent of the mass. The lenses are 0.3 to 0.6 m across and 0.1 to 0.3 m high. Laminar stromatoporoids are dominant, together with some tabulate and rugose corals. The main stromatoporoids are *Stromatopora*, *Actinostroma* and *Labechia*, the main corals *Favosites*, *Halysites* and *Heliolites*. The corals *Thecia* and *Alveolites* form laminar growths. Bryozoa and brachiopods occur towards the top of the reefs, and fragmentary debris within the micrite mass consists mainly of crinoid debris, along with brachiopod, bryozoan and molluscan fragments. Algae occur throughout; *Solenopora* occurs on the edges of the reefs, commonly associated with the stromatoporoids, *Rothpletzella* coats corals and is locally intergrown with them, and *Girvanella* coats shells and corals and occurs as isolated filaments.

LUDLOW SERIES

The Ludlow Series is represented by the Lower Ludlow Shale, the Aymestry Limestone Formation and the Upper Ludlow Shale. Table 2 shows the various classifications proposed for the series and their correlation with the one used in this account and that of the Ludlow type area.

Phipps and Reeve (1967) concluded that all the subdivisions of the Ludlow Series thin southwards from West Malvern towards the Gorsely axis (Lawson, 1954). They also suggested that superimposed on this trend is a local facies change and thickness increase in a trough-like depression centred on Bradlow [715 388], 1 km south of the district. However, because of the complex structure in this area, estimates of thickness are unreli-

Table 2 Classifications of the Ludlow Series in the Malverns area compared with that of the Ludlow, Shropshire type area.

MALVERNS AREA							LUDLOW TYPE AREA	
Phipps, 1957		Phipps and Reeve, 1967		Phipps and Reeve 1969	Penn, 1969	This Memoir	Holland et al., 1963	Holland et al. (1980); Lawson and White (1989)
Coddington Formation	Coddington Beds	Upper Ludlow Formation	Whitcliffe Flags Member	Upper Ludlow Shales	Upper Barton Court Beds	Upper Ludlow Shale	Upper Whitcliffe Beds	Upper Whitcliffe Formation
					Lower Barton Court Beds		Lower Whitcliffe Beds	Lower Whitcliffe Formation
Chance's Pitch Formation	Upper Chance's Pitch Beds		Woodbury Shale Member		Upper Godwins Rise Beds		Upper Leintwardine Beds	Upper Leintwardine Formation
	Lower Chance's Pitch Beds		Mocktree Shale Member		Lower Godwins Rise Beds		Lower Leintwardine Beds	Lower Leintwardine Formation
Hope End Formation	Dogg Hill Limestone	Aymestry Formation	Aymestry Limestone Member	Aymestry Limestone	Upper Halesend Beds	Aymestry Limestone Formation	Upper Bringewood Beds	Upper Bringewood Formation
	Rilbury Siltstones		Rilbury Siltstone Member	Lower Ludlow Shales	Lower Halesend Beds	Lower Ludlow Shale	Lower Bringewood Beds	Lower Bringewood Formation
Oldcastle Formation	Upper Oldcastle Beds	Lower Ludlow Formation	Oldcastle and Bradlow Facies		Upper Hill House Beds		Upper Elton Beds	Upper Elton Formation
	Lower Oldcastle Beds				Lower Hill House Beds		Middle Elton Beds	Middle Elton Formation
							Lower Elton Beds	Lower Elton Formation

able and the changes described by Phipps and Reeve cannot be confirmed.

As in the type area to the west of Ludlow, the rocks of the early part of the series yield a varied shelly macrofauna which gradually becomes less diverse upwards until there is a greatly restricted fauna of mainly brachiopods and molluscs in the uppermost beds, although fossils remain common. Graptolites are rare, but have been recorded by Penn (1969) at a few localities. Sections through the Aymestry Limestone and into the upper part of the Upper Ludlow Shale (equivalent to beds of the Lower Whitcliffe Formation) have been sampled for conodonts in the road cutting [7455 4019–7490 4024] on the north side of the A449 at Chance's Pitch and on the south-east side of Godwin's Rise, Frith Wood [7231 4024–7206 4010]. The collections comprise a conservative Ludlow fauna of *Ozarkodina confluens*, *O. excavata*, *Panderodus* spp., *Oulodus* sp. and, except in the lowest samples, *Coryssognathus dubius*. Rare specimens of *Kockelella* sp. occur in the lowest beds of the Upper Ludlow Shale (the Lower Leintwardine Formation). In a prominent limestone in the topmost Ludlow beds at Brockhill Quarry [7568 4394], the standard Ludlow fauna is joined by a relatively small number of specimens of *Ozarkodina* aff. *remscheidensis eosteinhornensis*.

Lower Ludlow Shale

The Lower Ludlow Shale comprises 220 to 240 m of silty mudstones and siltstones. They are generally calcareous; micritic limestone nodules and layers occur throughout, but are particularly common in the transitional beds at the top and base of the formation. Where fresh the mudstones and siltstones are blue-grey, but weather to olive-grey and buff and give rise to heavy silty clay soils on an outcrop that mainly lies in the poorly exposed low-lying ground between the ridges of the Much Wenlock and Aymestry limestones. Comminuted shell debris is commonly concentrated into thin layers. Bentonites are present throughout the formation, the thickest recorded (0.61 m) being at Barton Farm [744 401] (Phipps, 1957).

The formation represents all except the highest part of the Gorstian Stage. The lowest beds contain a diverse shelly fauna dominated by small brachiopods. Anthozoans and stromatoporoids, so common in the underlying Much Wenlock Limestone, are absent, but molluscs are well represented. From these early Gorstian beds Penn (1969) listed *Atrypa reticularis*, *Cyrtia exporrecta*, *Dicoelosia biloba*, *Eospirifer* sp., *Howellella elegans*, *Jonesea* [*Aegiria*] *grayi*, *Leptaena* (small variety), *Lingula* sp., orthids, *Protochonetes* cf. *minimus*, bivalves and *Dalmanites* sp., together with *Saetograptus (Colonograptus) varians*, associated with the dendroid *Callograptus*? The range of *Saetograptus varians* spans the *N. nilssoni* and *L. scanicus* biozones, the basal biozones of the Ludlow Series. A similar diverse fauna, with numerous small brachiopods characterises the Lower Elton Formation of the type area.

In higher beds, fossils are much less abundant, although the assemblage is similar to that of the lower beds, except that large strophomenids become more common. Also at this level the saetograptids *S. (C.) varians* and *S. (S.) chimaera s. l.* have been found (Penn, 1969), indicating a horizon within the *N. nilssoni–L. scanicus* biozones.

In the highest beds large brachiopods become numerous and molluscs, especially bivalves, increase in diversity and numbers. Solitary corals are also well represented. These beds are the Rilbury Siltstone Member of the Aymestry Limestone Formation of Phipps and Reeve (1967). Their faunal list includes *Amphistrophia funiculata*, *Atrypa reticularis*, *Coolinia pecten*, *Craniops implicatus*, *Gypidula* sp., *Howellella elegans*, *Isorthis orbicularis*, *Leptaena* sp., *Leptostrophia filosa*, *Microsphaeridiorhynchus nucula*, *Shagamella minor* [as *Chonetes lepisma*], *Sphaerirhynchia wilsoni*, *Strophonella euglypha*, *Cardiola interrupta*, *Cypricardinia subplanulata*, *Goniophora cymbaeformis*, *Poleumita globosa*, *Dalmanites* sp. and *Encrinurus* sp. This fauna indicates a correlation with the Lower Bringewood Formation of the type area, which is supported by Penn's (1969) record of a possible example of *Pristiograptus tumescens* at this level, probably indicating the *P. tumescens* Biozone of late Gorstian age.

The best sections seen during the survey were:

1. Temporary section [7335 5046] west of Bachelor's Bridge, Suckley exposing about 30 m of olive-green calcareous silty mudstones and siltstones with subordinate nodular layers of pale grey argillaceous limestone. Common fossils are *Atrypa reticularis*, *Eospirifer* sp., *Gypidula* sp., *Leptaena* sp., *Orthis* sp. and *Dalmanites myops*.

2. Storridge road cutting [7435 4835–7432 4828] exposing about 45 m of grey-green silty mudstones and siltstones. Thin limestones and layers of limestone nodules are present in the basal beds, and there are some thin bentonites in the higher beds. The topmost beds are exposed on the east side of a track opposite Brook House Farm and a further 4 m of siltstones and silty mudstones are exposed on the other side of the track entrance. All the beds are calcareous and fossiliferous. Bullard (1989, p.54) recorded the brachiopods *Atrypa*, *Dicoelosia*, *Howellella*, *Isorthis* and *Resserella*, and noted that trilobites, particularly *Dalmanites*, are fairly common. Penn (1969, p.149) listed a diverse shelly fauna, including solitary corals, bryozoans, brachiopods, molluscs and arthropods.

3. Excavation [7425 4040] at a barn near Kyre Cottage in 2 m of tough brownish grey poorly laminated siltstone with small limestone nodules. A late Gorstian fauna of *A. reticularis*, *Gypidula* sp., *L. depressa*, *S. euglypha*, *Poleumita globosa* and *Dalmanites myops* was collected.

Aymestry Limestone Formation

The Aymestry Limestone comprises 18 to 25 m of limestones and calcareous siltstones. The proportions of limestone and siltstone vary, but generally the narrow outcrop of the formation makes an easily mappable, bold feature. The precise top and bottom of the formation are less easily mapped, being transitional with the overlying and underlying beds.

The limestones range from grainstones to argillaceous wackestones. They are most commonly impure and nodular, and occur in mainly thin to medium beds, although thick beds also occur and some channel forms have been observed. They are blue-grey where fresh, but are commonly weathered to olive-grey. The calcareous siltstones are blue-grey to olive-grey, and show small-scale cross-bedding. Mudstone interbeds are common, and thin bentonites occur throughout.

The formation is relatively poorly fossiliferous; Phipps and Reeve (1967) noted that the fauna is similar to that of the underlying Rilbury Siltstone, and differences are a function of facies control and ecology. The fauna is essentially the same as that recorded from the upper part of the Lower Ludlow Shale except that corals are more common and include species of *Favosites*, *Heliolites* and *Syringopora*. Penn (1969) reported *S. (S.) chimaera s. l.* and unidentifiable graptolites, possibly indicating the *P. tumescens* Biozone. This supports a correlation with the Upper Bringewood Formation of the type area and indicates a late Gorstian age. However, the highest beds of the formation contain a lower Ludfordian Stage fauna at some localities (e.g. localities 1 and 2 below). This confirms Phipps and Reeve's (1967) observation that the upper part contains a fauna similar to that of their overlying Mocktree Shale Member. It is uncertain whether this is a persistent feature of the Aymestry Limestone in the present district, or is a local diachronism of the base of the Upper Ludlow Shale. The same relationship occurs at the type section for the base of the Ludfordian Stage at Sunnyhill Quarry, near Ludlow (Lawson and White, 1989).

The best sections are:

1. Disused quarry [7370 5592] near Knightsford Bridge, Knightwick. The sequence is overturned, and the beds are mostly fractured and tectonically disturbed, with some mineralisation in joints. Phillips (1848) showed a conformable sequence from these beds up into the Old Red Sandstone. This seems unlikely, as the Downton Castle Sandstone is not shown to be present. He recorded 15 m (50 feet) of Aymestry Limestone overlain by 18 m (60 feet) of Upper Ludlow beds. The beds exposed in the quarry are calcareous siltstones and silty limestones. The section reads:

	Thickness m
5. Siltstone, calcareous, olive-green, with silty, fine-grained limestone; shells throughout; massive in top 2 to 3 m	5.0
4. Siltstone, calcareous, olive-green, with thin limestone layers	13.0
3. Limestones, silty; pale grey; olive-grey, argillaceous in top few metres; rather disturbed; thin- to medium-bedded; shells throughout	8.0
2. Calcareous mudstones with silty limestones; thinly, irregularly bedded; shells throughout	1.3
1. Limestones, pale to medium grey, silty, in thin (5 cm) irregular beds; shells throughout	5.0

Beds 1 and 2 contain a band of *Shaleria ornatella*, beds 3 and 4 contain *Sphaerirhynchia wilsoni*, *Orbiculoidea rugata* and a leptostrophomenid?; *Microsphaeridiorhynchus nucula*, *Protochonetes ludloviensis* (abundant) and *?Isorthis* sp. occur in Bed 5. It would appear therefore that the Aymestry Limestone is here of early Ludfordian age.

2. Disused quarry [7320 5035] at Longley Green:

	Thickness m
Upper Ludlow Shale (p.35)	—
AYMESTRY LIMESTONE	
Siltstone, very calcareous, and argillaceous limestone, with thin beds of very shelly limestone; mid-grey to greenish grey, thinly bedded and sporadically laminated, becoming thinner bedded downwards with a weak nodular fabric locally; poorly fossiliferous overall, but with a few very fossiliferous beds; poorly exposed in upper part; *Protochonetes ludloviensis*, *Sphaerirhynchia wilsoni*, *Fuchsella amygdalina*, *Leiopteria* sp. and *Serpulites longissimus*	c.8.0
Limestone, argillaceous, and calcareous siltstone, mid-grey, fairly massive in part with subordinate nodular fabric and less common thin beds; *Microsphaeridiorhynchus nucula*, *Protochonetes ludloviensis*, *Sphaerirhynchia wilsoni*, *Leiopteria* sp., and *Serpulites longissimus*	3.0
Siltstone, very calcareous, and argillaceous limestone, with subordinate thin beds of shelly limestone, pale to mid-grey; mainly thinly bedded, with nodular fabric locally; some layers of shell debris	1.5
Siltstone, calcareous, with subordinate argillaceous limestone, slightly greenish grey, fairly massive with a weak nodular fabric locally; shelly locally	0.7
Limestone, argillaceous, and calcareous siltstone, mid-grey and green-grey; thinly bedded, with some cross-laminated beds; sporadically fossiliferous	0.9

The base of the formation lies close below.

3. Disused quarry [7378 4866], Halesend (Plate 4), exposes a section that ranges from the Aymestry Limestone into the Upper Ludlow Shale (p.35). The uppermost 7.5 m of the formation are exposed in the face, and the lower beds are patchily exposed to the east. The top of the formation is placed at the top of the 4 m of calcareous siltstones, which overlie 3.5 m of cleaner limestones. The top is marked by a prominent bedding plane, above which is a shaly, probably bentonitic, mudstone forming a recess (Bullard 1989). Penn (1969) recorded an erosion surface and thin limestone conglomerate at this level. The limestones are blue-grey, fine-grained, thinly parallel bedded and nodular-bedded, with shale partings.

4. Disused quarry [7406 4763] Vinesend Lane, exposes about 23 m of thinly bedded calcareous siltstones and parallel bedded to nodular limestones. Thin mudstone interbeds, one of which has a thin bentonite at its base, form recesses.

5. Disused quarry [7513 4569] near Mathon exposes 10 m of mainly thickly bedded limestones with buff shale interbeds. The limestones are blue-grey, mottled buff and fine-grained. Brachiopods are scattered throughout and locally concentrated on bedding planes. Some beds are of nodular limestone and a 1.5 m-thick bed near the base of the quarry has a channel form.

6. Bank section [7301 4181] near Old Colwall exposes beds from near the top of the Lower Ludlow Shale into the Aymestry Limestone:

	Thickness m
AYMESTRY LIMESTONE	
Siltstone, tough, calcareous, weakly laminated; abundant ovoid limestone nodules; common solitary corals and rare cf. *Spongarium aequistriatum*; *S. wilsoni*, *Poleumita globosa*	c.3

	Thickness m
Limestone; coalesced nodules in unlaminated siltstone	c.5

LOWER LUDLOW SHALE

Siltstone, tough, calcareous, flaggy, with abundant limestone nodules and some coquinal limestone beds; large strophomenids	c.4

7. Cutting [7258 4102–7254 4094] below Hope End House exposes about 6 m of the lower part of the formation. Poorly fossiliferous, tough, olive-grey, calcareous siltstones contain abundant limestone nodules coalesced into thin beds; the limestones occupy about 30 per cent of the rock. The basal 2 m are laminated, with *Favosites* sp., *Atrypa reticularis*, *Gypidula lata*, *S. euglypha* and *P. globosa*. Solitary corals, *Howellella elegans*, *L. depressa*, *P. ludloviensis* and *S. wilsoni* occur in the higher beds.

8. Disused quarry [7276 4093] opposite Lavanger Coppice exposing about 6m of poorly fossiliferous, massive, open-jointed limestone consisting of nodules coalesced into thin beds in a brown-grey to red-brown calcareous siltstone matrix.

9. Track exposures [7209 4004–7228 4021] on Godwins Rise in which up to 3 m of Aymestry Limestone are exposed (Penn and French, 1971, pp.20–21).

10. Disused quarry [7453 4027] north of Chance's Pitch exposing the highest part of the Aymestry Limestone; about 5 m of brown-grey tough calcareous poorly laminated siltstone contain numerous limestone nodules and layers; *A. reticularis*, *L. depressa*, *M. nucula*, *S. wilsoni*, *S. euglypha* and *D. myops* were collected.

Upper Ludlow Shale

The Upper Ludlow Shale consists mainly of olive-grey calcareous siltstones and mudstones. Thickness is generally 50 to 60 m, but increases to a maximum of about 100 m in the south around Coddington. Phillips (1848) recorded 18.3 m (60 feet) at Knightwick, but the section may be faulted (p.33). Limestone nodules and layers are

Plate 4   Ludlow Series rocks, Halesend Quarry [7378 4866]. Calcareous siltstones of the Upper Ludlow Shale overlie thinly bedded limestones of the Aymestry Limestone Formation. The junction is placed at the prominent bedding plane above the car bonnet. A 15081.

common in the lower part, particularly towards the base, in the transitional zone with the Aymestry Limestone. Thin limestone coquinas are present in these lower beds, and limestone conglomerates occur near the base. In the upper part the beds are flaggier, limestone nodule layers are less common and limestone coquinas are commonly weathered to brown rottenstone. Fine-grained, micaceous, silty, cross-stratified sandstones appear in the topmost beds. Bedding planes crowded with *Protochonetes ludloviensis* are a characteristic feature, and *Dayia navicula* is common towards the base of the formation. Phillips (1848) noted some layers, similar to the Ludlow Bone Bed, of white sandstone with black mica and fish fragments at Halesend (Locality 2).

The sequence of faunas present in the formation (Phipps and Reeve, 1967; Penn, 1969) closely matches the faunas of the Lower Leintwardine, Upper Leintwardine, Lower Whitcliffe and Upper Whitcliffe formations of the type area (Lawson and White, 1989).

The lowest beds are equivalent to the Lower Leintwardine Formation. The fauna is characteristic of the earliest Ludfordian and represents a sharp transition from the latest Gorstian. It includes taxa which are much commoner than in the older strata, such as *Dayia navicula*, *Howellella elegans*, *Hyattidina canalis*, *Isorthis orbicularis*, *Microsphaeridiorhynchus nucula*, and *Protochonetes ludloviensis*. Strophomenids are much reduced in numbers and diversity, although some, *Leptostrophia filosa* for example, remain common. *Salopina lunata* has not been recorded from lower beds, and there is a further increase in the diversity of the molluscan fauna.

In the higher beds the brachiopod *Shaleria ornatella* is abundant and associated with the trilobites *Calymene puellaris* [formerly *C. neointermedia*] and *Encrinurus stubblefieldi*, together with the ostracod *Neobeyrichia lauensis*, the occurrence of which is confined to this level. This assemblage is widely recorded in the Welsh Borderland and provides a firm link with the Upper Leintwardine Formation of the type area. Unidentifiable spinose monograptids recorded by Penn (1969) are likely to be *Saetograptus* (*S.*) *leintwardinensis*.

The uppermost beds of the formation correlate with the Lower and Upper Whitcliffe formations of the type area, with closely comparable faunas of latest Ludfordian age. Although brachiopods are numerous, there is a marked reduction in their diversity. The commonest forms are *Craniops implicatus*, *Howellella elegans*, *Microsphaeridiorhynchus nucula*, *Protochonetes ludloviensis* and *Salopina lunata*. There is no corresponding decrease in the molluscan fauna, with bivalves, gastropods and cephalopods all well represented.

Phipps and Reeve (1967) noted that the formation is characterised by features indicating wave and current action, and these become commoner higher up. They also noted the general increase in grain size with time culminating in the deposition of the overlying Downton Castle Sandstone. At least some of the coquinal limestone layers may be shell debris concentrated by storm activity.

The erosion surface at the base of the formation noted by Penn (1969) at Halesend (p.33) is absent to the south, but an intraformational conglomerate is present and a faunal change takes place (Cherns, 1980). Also, a plastic clay (possibly bentonite) is present at or close above the junction.

The following lists the best sections of the Upper Ludlow Shale:

1. Disused quarry [7320 5035], Longley Green, exposing 12.4 m of sparsely fossiliferous calcareous siltstones overlying the Aymestry Limestone (p.33). There are some shelly lenses, and *Atrypa reticularis* and *Salopina lunata* were noted.

2. Disused quarry [7378 4866], Halesend, exposing about 12 m of beds, and the boundary with the Aymestry Limestone (p.00). They mainly comprise calcareous siltstones, with some limestones, and thin shale interbeds. There are some fossiliferous coquinal limestones, and some bedding planes are crowded with brachiopods. The locality formerly afforded a complete section of the Upper Ludlow Shales, Phillips (1848) recording 46.6 m (153 feet) of beds.

3. Recently dug quarry [7400 4622] near Overley Cottages, Mathon, exposes about 25 m of siltstones with some shale layers. There are some limestone layers and shelly bedding planes; *Microsphaeridiorhynchus nucula* is abundant.

4. Brockhill Quarry [7565 4395] exposes the topmost beds of the formation and the overlying Ludlow Bone Bed and Downton Castle Sandstone (p.36). The Upper Ludlow Shale comprises mainly 13.7 m of calcareous siltstones with a few limestones layers and nodules overlain by 6 m of thinly bedded, micaceous, current-rippled sandstones with soft shale interbeds (Bullard, 1989; Penn, 1969, fig. 76). The siltstones contain *Protochonetes ludloviensis*, *M. nucula* and *Serpulites*; the topmost beds are poorly fossiliferous, containing some *Protochonetes* and *Lingula*.

5. Road cutting and extensive quarries [7243 4261–7249 4254] at the north end of Coombe Hill, Coddington exposing about 45 m of the top part of the formation, including beds with *S. ornatella*.

6. Roadside section [7200 4086–7202 4083] east of Loxeter through about 22 m of the lower part of the formation (Penn and French, 1971, p.21). Beds with *S. ornatella* are exposed at the top of the section.

7. Trackside section [7261 4098] comprising about 6 m of olive-grey siltstone with thin limestone lenses. *D. navicula*, *M.nucula*, *P. ludloviensis*, *S. lunata*, *Sphaerirhynchia wilsoni* and *Loxonema* were collected.

Fossils in the BGS collections from the topmost beds of the formation proved in the Colwall Borehole [7572 4265] were identified by C J Stubblefield and listed by Richardson (1935) as *Camarotoechia nucula*, *C.* cf. *nucula*, *Chonetes* sp., *Goniophora* aff. *cymbaeformis*, *G.* sp., *Nuculites* cf. *antiquus*, *Orthonota amygdalina* and *Orthoceras* sp.

PŘÍDOLÍ SERIES

The Přídolí Series comprises the rocks of Old Red Sandstone facies traditionally referred to as the Downtonian.

Its base, as with the base of the Old Red Sandstone, is placed at the base of the Downton Castle Sandstone Formation, the junction marked locally by the Ludlow Bone Bed Member, a thin lenticular phosphatic lag deposit that commonly occurs as infills of scour and ripple troughs in the topmost surface of the underlying Upper Ludlow Shale. The bed is recorded at only two localities within the district — at Halesend (Phillips, 1848; Reeve, 1953) and at Brockhill, from where it is locally named the Brockhill Bone Bed (see below).

The bone bed and base of the Downton Castle Sandstone were formerly regarded as marking the sudden change from shallow marine Ludlow sedimentation to the fluvial sedimentation of the Old Red Sandstone. However, more recently, the Downton Castle Sandstone has been interpreted as a littoral marine sand complex (e.g. Allen, 1985) and a storm-deposited shelf sand (Smith and Ainsworth, 1989). The bone bed is therefore more likely to have been a low-stand regressive lag deposit, later reworked and deposited in ripple troughs and scour hollows during a transgressive event (Ainsworth, 1991) rather than representing a simple regressive event (Richardson and Rasul, 1990, 1991). The mixing of brachiopods typical of the Ludlow Series with ostracods characteristic of the basal part of the Downton Castle Formation indicates that there is no significant time gap represented by the bone bed.

Most of the Přídolí is represented by the Raglan Mudstone Formation (Ledbury Formation of White and Lawson, 1989). The Raglan Mudstone is the lowest red-bed facies of the Old Red Sandstone and represents alluvial deposition on a coastal plain subject to repeated emergence and periodic transgressive influxes of shallow-marine to brackish water. The top of the Přídolí Series (and the Silurian–Devonian boundary) lies within this formation, but because of the lack of fossils its precise position is as yet unknown.

Downton Castle Sandstone Formation

The Downton Castle Sandstone is a flat-bedded sandstone, generally about 4 to 5 m thick. It may thin somewhat northwards in the Suckley–Knightwick area, and may be locally absent there (Phillips, 1848), but its absence is thought to be more likely due to faulting (p.33). It is variously coloured but mainly buff or yellow and locally pale grey to white, micaceous, and commonly feldspathic and speckled. Two samples from the Coddington area are pressure-welded quartz-rich sandstones with minor potassium feldspar and glauconite (Strong, 1986). Generally fine- to medium-grained, the sandstone contains thin siltstone interbeds, and plant debris is locally common. In some places a small feature marks its outcrop. The Ludlow Bone Bed at its base contains fish remains (mainly spines and thelodont scales). Phipps and Reeve (1967, p.353) reported rare examples of *Lingula lewisii*, *Microsphaeridiorhynchus nucula*, *Protochonetes ludloviensis* and *Salopina lunata* from the overlying beds. Examples of the ostracod *Londinia* cf. *fissurata* occur at this level, and *Frostiella groenvalliana* and *Londinia* sp. have been reported from the southern

Malverns (Worssam et al., 1989, p.14). These are common immediately above the Ludlow Bone Bed at Ludford Lane, Ludlow, and are of national and international value in correlating the base of the Přídolí Series (White and Lawson, 1989).

The formation is poorly exposed in an outcrop that is rarely more than 10m wide. The following are the better sections:

1. Small disused quarry [7402 4710], Church Stile Farm, Cradley, in orange and buff weathered fine-grained sandstone with burrows.

2. Stream exposures [7399 4689; 7400 4687; 7400 4676] near Cradley of tough, fine-grained, flaggy, parallel-laminated, orange-weathering sandstone about 3 m thick.

3. Disused quarry [7532 4526] near Mathon exposing 3 to 4 m of flaggy, parallel-bedded, pale grey to white, fine-grained, quartzitic sandstone weathering to buff and orange with some greenish grey micaceous silty partings.

4. Brockhill Quarry [6365 4395] (Phillips, 1848; Richardson, 1907; Stamp, 1923; Reeve, 1953; Penn, 1967, p.98; Penn and French, 1971, p.28). This disused quarry was overgrown during the survey in 1982, but has recently been cleared (Bullard, 1989, p.59). Stamp recorded the Brockhill (Ludlow) Bone Bed as a bed, 0.6 to 2.5 cm ($\frac{1}{4}$ to 1 inch) thick, of rolled phosphatic nodules and ferruginous concretions with fish spines and inarticulate brachiopods lying in hollows in the underlying Upper Ludlow Shale surface. Reeve noted that the bed consisted of fragmented remains of chitinous skeletons and phosphatic nodules. Penn recorded the bone bed as 7.6 to 15.4 cm (3 to 6 inches) thick, with evidence of channelling and burrowing immediately below; Penn and French recorded thelodont fragments. Bullard recorded small lenses of sandstone with fish remains and phosphatic material. The overlying beds are buff, fine-grained, micaceous sandstones, sandy shales and shales with plants.

5. Section [7297 4190] behind Old Colwall House of 0.6 m of olive-grey, flaggy, finely sandy siltstone with indeterminate gastropod and bivalve fragments, the ostracods ?*Cytherellina siliqua* and *Londinia* cf. *fissurata*, the alga *Pachytheca* and other plant fragments.

6. Section [7126 4032] in bank at the Old School House comprising about 2 m of pale brown, fine- to medium-grained sandstone in beds 2 to 5 cm thick.

7. Exposure [7205 4171] by a forestry track in Berrington Wood of the basal beds of the formation, comprising 0.6 m of brown, laminated, sandy siltstone and fine-grained, silty sandstone with *P. ludloviensis*, *S. lunata* and *Lingula* cf. *minima*. Temporary sections [7187 4141; 7179 4127] were recorded along strike to the south and a small exposure noted [7175 4118].

The formation was proved in the Colwall Borehole [7572 4265] where it was about 4.7 m thick. A sample in the BGS collection is medium-grained, pale greyish buff, micaceous sandstone.

Raglan Mudstone Formation

The Raglan Mudstone Formation is almost entirely of red-bed facies. The Silurian–Devonian boundary occurs within the formation, but, for convenience, the whole of the formation is described in this chapter. About 800 m are estimated to be present in the district. The Collington Borehole [646 610], close to the north-west of the district, proved the basal 560 m (Department of Energy, 1978; Tunbridge, 1983).

The formation has a wide outcrop in the west of the district. It consists predominantly of red mudstones and siltstones with calcretes. Sandstones occur throughout; a local ferruginous sandstone facies is present at the base of the formation in the Wellington Heath area and a local pebbly quartzitic sandstone facies is present towards the top of the formation in the Bringsty Common area. Apart from these areas, the soft nature of the rocks produces a subdued topography in which there are few good sections.

Heavy mineral analyses of some sandstones from the Wellington Heath beds show differences from those of sandstones of the Raglan Mudstone (and overlying St Maughans Formation) from the adjoining Hereford and Ross-on-Wye districts (Brandon, 1989; Morton, 1988). Garnet is, as in the other mineral suites, the most abundant member, with a mean value of 53.2 per cent and a range from 29 per cent to 76 per cent. However, unlike the other suites, tourmaline is also abundant, with a mean of 32.5 per cent and a range of 18 per cent to 51 per cent. Rutile is also considerably enriched compared to the other suites (mean 9.5 per cent) and staurolite is more conspicuous. Zircon and apatite contents are low (mean 2.9 per cent and 1.4 per cent respectively). Epidote is a relatively consistent but minor phase, and calcic amphibole was observed in one sample. Diagenetic carbonate is present in one sample (E60157), apparently preferentially replacing garnet.

The results are similar to those of a study by Allen (1974a) of the Downton Castle Sandstone, Ledbury Formation (Raglan Mudstone Formation), and Ditton Series (St Maughans Formation) of the Clee Hills, apart from the fact that Allen treated his samples with acid and dissolved the apatite. Also, the present analyses are of the 63 to 125 μm fraction, whereas Allen's analyses were apparently of the whole-rock suite. The similarity of the suites of the Downton Castle Sandstone and the Wellington Heath beds may indicate a common provenance, but further work on the Downton Castle Sandstone within the district is required.

The extremely high garnet contents of the Raglan Mudstone Formation indicate a provenance dominated by metamorphic basement, presumed to be the newly uplifted Caledonian mountain belt to the north (Allen, 1974b; Allen and Crowley, 1983). The appearance of zircon, rutile and tourmaline that Allen noted higher in the Old Red Sandstone was ascribed by him to the increasing input from more proximal sedimentary and volcanic rock sources as the Caledonian front advanced southward. However, the most significant trend shown by the present analyses is the upward increase in apatite, which is primarily of igneous origin.

The provenance of the Wellington Heath beds is difficult to assess. Their high tourmaline content is of significance, but without geochemical investigation it is not clear whether it is of igneous (pegmatitic, as suggested by Allen, 1974a) or metamorphic (psammitic) origin. Reworking of local sedimentary rocks appears unlikely, as Fleet (1925) has shown that Cambrian, Ordovician and Silurian sandstones all contain more abundant zircon than either tourmaline or rutile. It is possible that some of the suite, particularly the garnet, is of similar derivation to the Raglan Mudstone suite and was introduced by axial drainage systems, whereas the tourmaline was derived from a laterally fed system.

The presence of epidote in the mineral suite suggests a relatively shallow burial history, perhaps of less than 1 km. It is unstable and suffers dissolution at low pore fluid pressures and is entirely dissolved at burial depths greater than 1100 m in Palaeocene sediments in the North Sea (Morton, 1984).

Details of the principal lithologies of the Raglan Mudstone Formation are given by Brandon (1989). They represent deposition in a coastal alluvial floodplain environment (Allen, 1985) with cyclic deposition of sands, silts and muds. The sandstones are channel deposits and the mudstones and siltstones are overbank floodplain deposits. Calcrete is abundant, mainly in the form of small limestone nodules which occur in the mudstones in the upper part of the cycles. Thick mature calcretes are also present and the Bishop's Frome (Psammosteus) Limestone at the top of the formation is a sequence of persistent, stacked mature calcretes up to about 8 m thick (Plate 5). Lingulids and modiolopsid bivalves occur locally at the base of the formation and point to marine flooding events when the coastal plain was covered by shallow marine to brackish water. Similar conditions prevailed during deposition of the basal part of the overlying St Maughans Formation (Barclay et al., 1994).

The Townsend Tuff Bed, a volcanic ash complex, lies about 100 m below the top of the Raglan Mudstone and crops out in the south-west of the district. It is present in two stream sections [70064 44520; 70145 44780] in an easterly tributary of Dowding's Brook (Figure 18). It comprises three falls (Allen and Williams, 1981a; Parker et al., 1983) throughout Wales and the Welsh Borderland, but only the upper two are exposed at these localities. The lowest exposed part of the complex, part of Fall B, is 0.04 m of micaceous crystal tuff, seen at the second locality. It is overlain by soft, pink, unctuous dust tuff, in turn overlain by pale green and red porcellanite. The top of Fall B is weathered to deep ochreous yellow. Fall C is 0.81 m thick and consists mainly of soft, unctuous, pink, clay-grade dust tuff, the basal 0.02 to 0.05 m of which are deep carmine red. The top of Fall C is marked by an immature calcrete. Volcaniclastic clays and silts occur at a similar level at three further localities in the district and are tentatively correlated with the Townsend Tuff. In the bank of Cradley Brook [7316 4872] north of Stifford's Bridge, dull green-grey, white and red sticky clays are exposed; green, white and red mottled clays are exposed in the bank [6852 5641] of Paradise Brook, Brockhampton; farther downstream [7028 5721–7033 5720] in Linceter

Plate 5
Bishop's Frome
Limestone,
Stanley Hill,
Bosbury [6758
4394]. The
limestone is a
mature, rubbly
calcrete.

Brook grey-green to grey-blue hard tuffaceous siltstone and red and green mottled bentonitic mudstones are exposed. Smectite is the dominant clay mineral of the complex, illite occurring in the overlying siltstones. The area lies in the field of clay mineral assemblage I of Parker et al. (1983, fig.4). These authors infer this to have had a relatively shallow burial history, which is supported by the presence of epidote in the heavy mineral suites (p.37).

The Bishop's Frome (Psammosteus) Limestone marks an important regional tectonic event (Allen, 1985), during which dispersal systems bypassed the Welsh Borderland and thick, mature calcretes formed on a surface that remained emergent for perhaps tens of thousands of years. There are some indications within the district that the event may be marked locally by an unconformity. In the Brockhampton area [690 570], the beds of the underlying Raglan Mudstone appear to be more folded than the Bishop's Frome Limestone and overlying St Maughans Formation.

The Bishop's Frome Limestone is partly exposed at the following localities:

1. Exposure [7250 5925] above Lower Tedney Farm, Whitbourne, of 5 m of red, blue and green mottled calcreted mudstones with columnar, prismatic pedogenic structure.

2. Former quarry exposure [6778 4848] on Ward Hill, Halmond's Frome, of 2 m of rubbly nodular to massive calcrete.

3. Recent excavation at Little Lapp Cottage [6751 4818], Halmond's Frome, of over 4 m of calcrete dip-

ping about 25° and consisting of two palaeosol profiles separated by a mature, massive calcrete.

4. Exposures in a ditch [6740 4558] near Hill Farm, Castle Frome; the limestone forms an extensive east-dipping slope strewn with limestone nodules in this area.

5. Exposures of mature calcrete in disused pit [6960 4465] north of Harbour Hill, on a ridge formed by the limestone that extends from here north-east to Beacon Hill.

The following lists the better sections of the Raglan Mudstone Formation below the Bishop's Frome Limestone:

1. Fine river cliff section [7340 5830] on the Teme exposing a sequence of about 12 m of channelised sandstones and calcreted red mudstones. Fish fragments are present in the base of a green, micaceous, medium-grained sandstone. Steeply dipping beds are well exposed [7355 5825] at river level to the east.

2. Fine river bank section [7290 5652] on the Teme near Ankerdine Farm exposing about 9 m of red mudstones with green layers and a lenticular sandstone near the base.

3. Disused quarry [7113 4106] near Wellington Heath exposing 3 m of purple-red-brown, fine- to medium-grained, micaceous, parallel-laminated and cross-laminated sandstone; some circular burrows, 4 mm in diameter. This is one of several quarries in the vicinity exposing this local facies at the base of the Raglan Mudstone Formation.

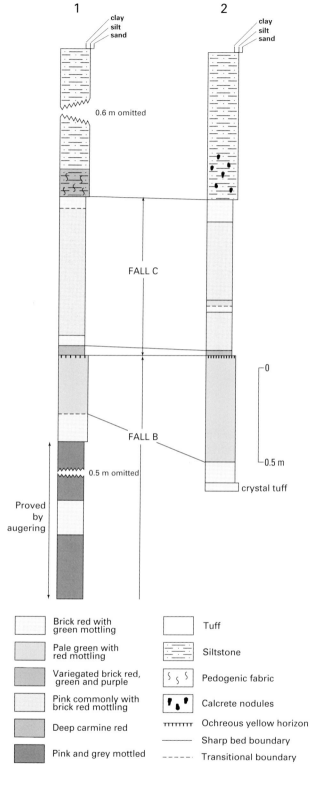

Brick red with green mottling

Pale green with red mottling

Variegated brick red, green and purple

Pink commonly with brick red mottling

Deep carmine red

Pink and grey mottled

Tuff

Siltstone

Pedogenic fabric

Calcrete nodules

Ochreous yellow horizon

Sharp bed boundary

Transitional boundary

The basal 353.56 m of the formation were proved in the Colwall Borehole [7572 4265] (Richardson, 1935), consisting of mudstones and sandstones. *Lingula* occurs in the lowest beds. The formation was recorded by Robertson (1926) in the adjacent railway cutting west of the Colwall Tunnel, faulted against the Coalbrookdale Formation. Close to the fault are purplish red-green, mottled mudstones and minor micaceous sandstones. A sequence of sandstones 183 m from the tunnel mouth (see Robertson, 1926, p.172 for details of section) contains a fish-bearing bed from which *Auchenaspis egertoni*, *Cephalaspis* and *Eukeraspis* were identified by W W King. Soft mudstones lie above the sandstones.

Figure 18 Sections of the Townsend Tuff Bed.
1 Section [7006 4452] in Dowding's Brook
2 Section [7014 4478] in Dowding's Brook.

FIVE

Devonian

The Devonian rocks belong to the Old Red Sandstone magnafacies of Wales and the Welsh Borderland. The basal part of this red-bed sequence, the Raglan Mudstone Formation, is mainly, if not all, of Silurian age and described in Chapter Four. The St Maughans Formation represents the Devonian in the district and belongs to the Lower Old Red Sandstone. It is of Lower Devonian age, spanning the Lochkovian (Gedinnian) Stage, although there is no evidence at present for the age of the topmost beds which may extend into the Pragian (Siegenian). Table 3 shows the classification of the Old Red Sandstone and the lithostratigraphical correlation with adjoining areas.

ST MAUGHANS FORMATION

The St Maughans Formation crops out in the higher ground in the west and north-west of the district, forming a plateau area fronted by a bold escarpment.

It consists of about 750 m of mudstones, sandstones and intraformational conglomerates. Overall, the mudstone to sandstone ratio is about 3 to 1, although the proportion of sandstones increases upwards, and the topmost beds in the district are sandstone dominated. Most of the sequence is identical to that of the adjoining Hereford district which was described in detail by Brandon (1989) and only a summary is presented here.

The mudstones are red-brown, purple and green. They range from being finely parallel-laminated to blocky and structureless. The structureless beds are generally palaeosols in which the original lamination has been destroyed by penecontemporaneous pedogenic processes. Root turbation and bioturbation are present, but calcrete formation and desiccation are the main processes to have affected the sediments. The calcrete occurs mainly in the form of small limestone nodules ('race'), but more mature massive to rubbly calcretes also occur, particularly towards the base of the formation. Body fossils are rare, apart from plant and eurypterid fragments in the green beds; bivalve–ostracod faunas present near the base of the formation at Ammons Hill (King, 1934; Barclay et al., 1994) have not been recorded elsewhere at this level. Few fish remains have been noted from the mudstones.

The sandstones are predominantly red-brown to dusky brown, but olive-green, purple-green and grey-green colours are also common, and the highest beds are mainly buff-yellow and olive-green. They are mainly fine to medium grained and well sorted, consisting of quartz, some feldspar and lithic grains with a calcareous cement. Mica is common, and, on some bedding planes, abun-

Table 3 Classification of the Old Red Sandstone rocks.

Lithostratigraphy			Biostratigraphy			Chronostratigraphy		
Ludlow/ West Midlands	This memoir		Fish zones	Thelodont zone	Miospore zones	Stages/series/ system		Local stages
						Pragian (Siegenian)	Devonian	Dittonian
Ditton Series	St Maughans Formation		*Pteraspis crouchi*		*Emphanisporites micrornatus – Streelispora newportensis*	— ? — ? — Lochkovian (Gedinnian)	Devonian	Dittonian
'*Psammosteus*' Limestones Group			*Pteraspis leathensis*			? — ? —	?	
Ledbury Formation	Bishop's Frome Limestone Member	Downton Group	*Traquairaspis symondsi*	*Turinia pagei*	*Synorisporites tripapilatus – Apiculiretusispora spicula*		Silurian	Downtonian
	Raglan Mudstone Formation	Downton Group	*Traquairaspis pococki*	*Logania kummerowi* + *Goniporus* + *Katoporus* spp.		Přídolí (Silurian)	Silurian	Downtonian
Temeside Shale Formation	Raglan Mudstone Formation	Downton Group					Silurian	Downtonian
Downton Castle Sandstone Formation		Downton Group					Silurian	Downtonian

dant. The heavy mineral suite is dominated by garnet (mean 70 per cent), although this proportion is less than in the underlying Raglan Mudstone (Allen, 1974a; Allen and Crowley, 1983; Morton, 1988). Apatite (mean 14.5 per cent) and rutile (mean 2.5 per cent), zircon (mean 4.6 per cent) are the other main constituents and chromite occurs in minor amounts. Opaque minerals are common and baryte is a common authigenic phase (Morton, 1988).

Large-scale trough cross-stratification, low-angle planar cross-stratification, parallel-lamination and small-scale rippled lamination are common. There are few sufficiently extensive sections to reveal the presence of epsilon cross-stratified laterally accreted sandstone bodies which are known to occur elsewhere in the formation. One or two examples of deformed cross-bedding have been recorded. Bottom structures on sandstones, parting lineations and cross-bedding generally point to a northerly source for the sediments (Allen and Crowley, 1983). The sandstones overlie erosional surfaces cut in the underlying mudstone or siltstone (Plate 6), and have intraformational debris in their basal parts.

The formation is characterised by the presence of calcareous intraformational conglomerates, the 'conglomeratic cornstones' of the older literature. Type A intraformational conglomerates (Allen and Williams, 1979) are the most common and occur as lenticular, channel-lag, polymictic deposits consisting of an unsorted mixture of calcrete, siltstone and mudstone clasts in a sand or silt matrix with a carbonate cement. They have low-angle cross-stratification and range up to 2 m in thickness. They occur mainly below the sandstones, but are also interbedded with them. Fish fragments occur in the conglomerates, and, to a minor extent, in the sandstones. The trace fossil *Beaconites antarcticus* (Allen and Williams, 1981b) occurs in a sandstone at Linton Tile Works (Plate 7).

The fining-upwards cycles described by Allen (e.g. 1964) from the formation elsewhere in the region, and interpreted as the deposits of laterally migrating meandering streams, are common. However, many of the sandstones are multistorey bodies comprising several stacked sandstone units separated by reactivation surfaces and commonly truncated by upper erosional sur-

Plate 6 St Maughans Formation, Linton Tile Works [668 538]. Channelised sandstones and overbank floodplain mudstones. A 13969

faces, suggesting a more proximal fluvial environment and perhaps a more braided stream system in the district, compared to the areas described by Allen.

The sequence was deposited as part of a southerly-prograding alluvial plain complex (Allen and Williams, 1979; Bluck et al., 1992), although the evidence from the Ammons Hill section (Barclay et al., 1994) points to marine influence during deposition of the basal beds.

Biostratigraphy

The biostratigraphical classifications of the Devonian rocks are shown in Table 3, although little data are available from the beds within the district.

The precise locations of early vertebrate discoveries are not known. Symonds (1872) recorded a collection of fossils from Acton Beauchamp, Bromyard and Stifford's Bridge, as well as from Ledbury to the south of the district. Lankester (1868) recorded *Pteraspis ?crouchii* from Cradley and Acton Beauchamp, and noted that *Pteraspis rostratus* occurs abundantly at Cradley along with *Scaphaspis lloydii*. Woodward (1891) catalogued the fish fragments from the district which are held in the British Museum (Natural History). He reinterpreted the *Pteraspis ?crouchii* of Lankester as *P. rostrata* and the *Scaphaspis lloydii* as probably part of *P. rostrata*. Also listed from Cradley are *Cephalaspis lyelli* and *C. salweyi*. Richardson (1907) summarised the early vertebrate discoveries. Stensiö (1932) described the following from Cradley: *C.salweyi, C. agassizi, C. fletti* (possibly from Cradley), *C. cradleyensis, C. langi* (possibly from Cradley), *C. sollasi, Benneviaspis lankesteri, B. anglica* and *B.* sp. *C. salweyi* was also collected from Acton Beauchamp. The Cradley specimens were described as apparently coming from the pits near Ridgeway Farm, which are in King's (1934) Zone II.1, and not from the large quarry in Zone I.9 at Ridgeway Cross, a mile to the south-east.

The fauna of the Ammons Hill section (Figure 19; Locality 10; King, 1934) is discussed in detail in Barclay et al. (1994). King recorded the fish fragments *Pteraspis leathensis* and *Didymaspis grindrodi*; Barclay et al. collected acanthodian spines of climatiid type. Bivalves include *Carditomantea virelyi, C reecei, Eurymyella ammonis, E.* cf.

Plate 7 Sandstone bedding plane with the trace fossil *Beaconites antarcticus*. St Maughans Formation, Linton Tile Works [6677 5378]. A 14155

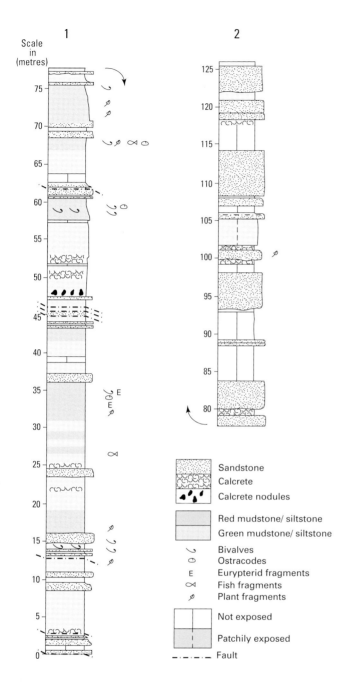

Figure 19 Section of the St Maughans Formation, Ammons Hill railway cutting. Section 1 is of trench section in the east (Barclay et al., 1994). Section 2 is of outcrops to the west [7000 5289–7006 5291].

shaleri var. *longa*, *Modiolopsis* cf. *complanata trimpleyensis*, *M. gradata*, *M. leightoni* var. *curta*, *M. nilssoni* and *M.* sp. The ostracod *Leperditia* sp. occurs in abundance in some beds; eurypterid fragments include *Pterygotus anglicus*.

Sections and exposures

There are numerous stream sections in the formation, although most are heavily vegetated and virtually impene-

trable between spring and late autumn. The sandstones commonly form steps and waterfalls, but the hardness of the groundwater emanating from the beds, and of the water in the streams, results in many outcrops being tufa coated. There are many disused quarries exposing mainly sandstone that was formerly dug for building stone. The Ammons Hill section (Locality 10; Figure 19; Barclay et al., 1994) provides the best record of the basal part of the formation. The following lists some of the better sections.

1. Disused quarry [6666 5735] in intraformational conglomerate on a closure of the Collington Anticline, 1 km south of Edvin Loach.

2. Stream exposures [6885 5960–6913 5962] near the base of the formation north of Tedstone Delamere.

3. Stream exposures [6812 5647–6830 5646] near the base of the formation in Paradise Brook, Brockhampton.

4. Disused quarry [6748 5674] near Sandy Cross exposing green and buff sandstones. They are channelised and comprise fining-upwards units with intraformational calcrete and mudstone clasts at their bases. A few exotic quartz pebbles also occur and there are also some intraformational conglomerate lenses. These beds resemble the Rowley Brook Beds of the Droitwich district (Mitchell et al., 1962, p. 52) and the Senni Beds of South Wales.

5. Disused quarry [6682 5540], Bromyard Downs.

6. Stream and waterfall exposures [7000 5990–7018 5998; 7043 5978; 7050 5938] in Sapey Brook and its tributaries.

7. Disused quarry [6807 5444], Brockhampton.

8. Disused quarry [6676 5379], Linton Tile Works, Bromyard (Plate 6). There are good exposures of the principal facies of the formation, although the quarry is flooded and access to many parts is not possible. About 24 m of beds are exposed, comprising calcreted mudstones and channelised sandstones. *Beaconites antarcticus* occurs in the top of the basal sandstone exposed (Plate 7).

9. Section [6685 5152] in bank of River Frome in which green mudstones with *Modiolopsis* sp. are exposed. These beds probably equate with the fossiliferous beds at Ammons Hill (Locality 10).

10. Disused railway cutting [7000 5287–7020 5300], Ammons Hill. Figure 19 is a graphic section of the beds exposed in the cutting and of those revealed by trenching of the eastern part of the cutting (Barclay et al., 1994).

11. Exposures [7000 5272–7095 5149] in stream south of Ammons Hill railway cutting. *Modiolopsis* sp. occurs in green silty mudstone at one locality [7014 5259]. The best section [7011 5229] exposes sandstones and intraformational calcrete-clast conglomerates.

12 Disused quarries [7103 5120; 7104 5109] near Suckley Court, mainly in intraformational conglomerates. The section in the southerly quarry is faulted.

13. Disused quarry [6884 4898] below Sinton's End Farm exposing up to 0.5 m of intraformational

conglomerate on up to 1.5 m of flaggy, cross-bedded, fine-grained sandstone.

14. Waterfall sections [6883 4873; 6898 4889] in a tributary of the Leadon near Evesbatch.

15. Intermittent stream exposures [6881 4778–6918 4688] in the Leadon below Evesbatch, amongst which one locality [6881 4778] provides easily accessible sections typical of the formation. They comprise cross-bedded channelised sandstones with intervening mudstones.

16. Large disused quarry [6739 46790] north of Fromes Hill.

17. Roadside sandstone exposures [6703 4637] on the A4103 west of Fromes Hill of sandstones; one bed of silty sandstone has deformed lamination.

18. Disused quarry [7085 4922] near Broom Farm in cross-bedded sandstones.

19. Large disused quarry [7187 4879] in pale green sandstones.

20. Large disused quarry [7165 4800] north of Pullen's Farm in channelised sandstones with an intervening red-brown mudstone.

21. Exposures [7309 4808] in Cradley Brook above Stifford's Bridge at the base of the formation, including green bentonitic, burrowed mudstones with a thin bentonite layer at the top.

22. Disused quarry [7023 4731], Upper Penshill Coppice, exposing sandstones with plant fragments above intraformational conglomerates.

23. Disused quarry [7156 4760] above Wells Farm exposing up to 8 m of channelised sandstones with mudstone interbeds.

24. Disused quarry [7184 4752] near Ridgeway House exposing interbedded lenticular sandstones and mudstones.

25. Large disused quarry [7210 4760] north of Ridgeway Cross exposing 2 m of mudstones above 2 m of speckled sandstones.

26. Roadside section [7151 4695], Tanhouse Farm, exposing sandstones and intraformational conglomerates.

27. Excellent disused quarry section [7166 4660] in intraformational conglomerates with minor sandstones.

28. Exposures in Cradley Brook [7314 4689–7314 4643; 7319 4603–7354 4595] mainly of sandstones, but some mudstones and intraformational conglomerates are also seen.

SIX

Carboniferous

Carboniferous rocks are confined to two small occurrences of Westphalian coal-bearing measures correlated with the Highley Formation of south Staffordshire (Murchison, 1839; Groom, 1900; Phipps and Reeve, 1967, 1969; Mitchell et al., 1962; Earp and Hains, 1971).

The first occurrence is at Berrow Hill [7435 5855], where up to about 25 m of beds unconformably overlie the Raglan Mudstone Formation, are unconformably overlain by the Haffield Breccia, and are truncated to the east by the East Malvern Fault. There are no sections, but greenish grey and yellow mottled clays are present at outcrop, and carbonaceous clay spoil occurs around old pits [7423 5827] at the south of the outcrop. Murchison (1839, p.135) referred to these old workings and Phillips (1848) recorded that some working had taken place thirty years prior to his survey. There are two smaller outliers immediately to the north of Berrow Hill, north of the district boundary, at Martley. The more northerly one surrounds the Precambrian/Cambrian inlier at Martley Pit (Mitchell et al., 1962).

These beds were named the Martley Measures by Wills (1956). They were referred to the Highley Beds by Mitchell et al. (1962), equivalent to the Halesowen Beds of south Staffordshire. Their equivalence to the Enville Beds was suggested by Wills (1956), but miospore evidence (Spinner, 1966) supports correlation with the Highley Beds (Mitchell et al., 1962, second impression, 1976) and indicates a probable Westphalian D age. They were renamed the Highley Formation by Ramsbottom et al. (1978).

The second occurrence is near the New Inn, Storridge [7595 4926], where there is a small exposure of grey carbonaceous mudstone degraded to clay in the stream below the A4103. This may be the locality referred to by Murchison (1839, p.135) as 'near Old Storridge Hill',

where former workings were reported to him. Groom (1910) referred to the locality as the southern end of Storridge Hill. These beds are part of a very small fault-bound sliver in the East Malvern Fault zone. The fault is exposed 170 m to the north [7693 4943] in a small stream, juxtaposing green Silurian mudstones against red Triassic mudstones (p.104) with no intervening Coal Measures (as reported by Phipps and Reeve, 1969, p.28). Similarly, they are absent in the Crumpton Hill Road 200 m to the south, where the fault can be located between purple and green Silurian mudstones to the west and red Triassic mudstones to the east.

No Coal Measures are known to be present at depth within the district. Some may occur adjacent to the East Malvern Fault. The prediction of Falcon and Tarrant (1951), Cook and Thirlaway (1955) and Phipps and Reeve (1969) that they might be present below the Permo-Triassic rocks of the Worcester Basin was disproved by the Kempsey Borehole (p.13), at least in the area of the basin within this district. Variscan inversion and erosion of the tract between the East Malvern and Inkberrow faults stripped the Palaeozoic rocks in this area, apart perhaps from remnants preserved close to these faults. The BGS Stratford seismic line shows that Carboniferous rocks are likely to be present to the east of an unnamed fault west of the Inkberrow Fault.

There is no evidence of the age of the Haffield Breccia, which unconformably overlies the measures on Berrow Hill and occurs as small outliers to the south (p.48). This was named the Clent Breccia by Mitchell et al. (1962) and assigned to the Carboniferous. It is here assigned a Permian age, in accord with Smith et al. (1974) and Worssam et al. (1989), although it may range from late Carboniferous to early Permian, and is described in Chapter 7.

SEVEN

Permian and Triassic

About three-quarters of the district is underlain by Permian and Triassic rocks which infill the Worcester Basin. The sequence comprises the Permian Haffield Breccia and Bridgnorth Sandstone Formation and the Triassic Sherwood Sandstone, Mercia Mudstone and Penarth groups. Most of the beds lie east of the East Malvern Fault, faulted against Precambrian and Palaeozoic rocks, but a few outliers rest unconformably on Palaeozoic rocks immediately west of the fault. In the centre of the basin, the Bridgnorth Sandstone rests unconformably on volcaniclastic rocks of probable Precambrian age. To the east of the district, seismic reflection evidence suggests the presence of Palaeozoic rocks beneath the Permo-Triassic rocks (Chadwick and Smith, 1988). In the east, Jurassic rocks conformably overlie the Triassic.

The Kempsey Borehole [8609 4933] (Figure 20) proved 2298.4 m of Permo-Triassic rocks, commencing in the Mercia Mudstone Group at a level just above the Arden Sandstone. Evidence from other boreholes and surface mapping suggests that up to 200 m of Triassic strata overlie the Arden Sandstone, giving a maximum thickness of about 2500 m of Permo-Triassic rocks in this part of the basin. Elsewhere, evidence from seismic reflection lines suggests considerable thickness variations across the district (Figure 39), with a maximum thickness of over 3000 m in the north-east.

The succession of the Permian and Triassic rocks is shown in Table 4. The earliest Permian sediments, the Haffield Breccia, were marginal breccias of fluvial and debris flow origin deposited in an arid to semi-arid continental climate. Deposition of these may have continued into the Triassic. Aeolian deposition of the sands of the Bridgnorth Sandstone Formation was followed by fluvial deposition of the sandstones and conglomerates of the Sherwood Sandstone Group. The dominantly fine-grained, windblown, red sediments of the succeeding Mercia Mudstone Group were deposited, together with the evaporites halite, gypsum and anhydrite, in sabkha to playa lake environments. A northward connection to an open-marine environment is indicated by the thick sequence of halite-bearing beds comprising the Droitwich Halite Formation to the north of the district. These beds probably extend across most of the eastern part of the Worcester district, east of the Smite, Pirton and Tewkesbury fault system (Figure 32). Generally arid conditions were interrupted by a humid, pluvial period, when the green beds of the Arden Sandstone Formation were deposited in brackish environments. Sea-level rise and marine transgression towards the end of the Triassic Period resulted in deposition of shallow-marine mudstones, sandstones and limestones of the Penarth and Lias groups.

The Permian–Triassic boundary is arbitrarily placed at the top of the Bridgnorth Sandstone. The marginal brec-

cias on the western edge of the basin, to the west of the East Malvern Fault, are correlated with the Haffield Breccia and assigned a Permian age, those to the east are included in the Triassic Sherwood Sandstone Group. The Triassic–Jurassic boundary is currently placed at the incoming of the ammonite *Psiloceras* (Cope et al., 1980a; Warrington et al., 1980). In the Worcester district, the junction is in the Saltford Shale Member in the north and in the Wilmcote Limestone Member in the south, near to the base of the Blue Lias Formation of the Lias Group. These rocks are described in Chapter 8.

Haffield Breccia

The term Haffield Conglomerate was introduced by Phillips (1848) and superseded by Haffield Breccia (Groom, 1902). It is analogous with the Clent Breccia of the area to the north (Mitchell et al., 1962). The name is applied to coarse, poorly sorted breccias and conglomerates which, in the Worcester district, occur in isolated outcrops immediately west of the East Malvern Fault and

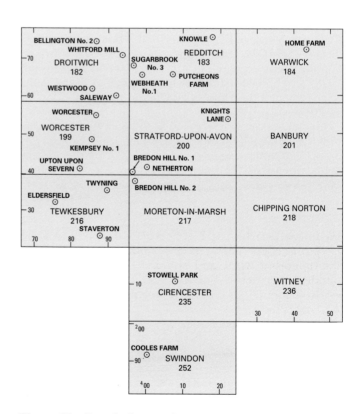

Figure 20   Boreholes proving Permo-Triassic and Jurassic strata mentioned in Chapters 7 and 8. BGS 1:50 000 sheets also shown.

Table 4
Lithostratigraphy and chronostratigraphy of the Permian and Triassic rocks.

Chronostratigraphy			Lithostratigraphy		
System	Series	Stage			
Triassic	Upper	Rhaetian	Lias Group*	Blue Lias Formation	Saltford Shale Member*
					Wilmcote Limestone Member
			Penarth Group	Lilstock Formation	Cotham Member
				Westbury Formation	
		Norian	Mercia Mudstone Group	Blue Anchor Formation	
				Twyning Formation	
		Carnian		Arden Sandstone Formation	
				Eldersfield Mudstone Formation	Droitwich Halite Member
	Middle	Ladinian			
		Anisian			Holling Member
			Sherwood Sandstone Group	Bromsgrove Sandstone Formation	Finstall Member
					Burcot Member
	Scythian			Wildmoor Sandstone Formation	
				Kidderminster Formation	
Permian				Bridgnorth Sandstone Formation	
				Haffield Breccia†	

* See Chapter 8
† Age uncertain (see text)

rest unconformably on Llandovery, Wenlock and Carboniferous rocks. The breccias include both matrix- and clast-supported types, and comprise mostly angular to rounded clasts up to boulder size, set in a red-brown, poorly cemented and friable, hematitic, variably muddy and sandy matrix.

Ramsey (1855) and Lapworth (1898) described these breccias throughout the Midlands and commented on their similar lithologies at all localities northwards from the South Malverns area. They noted the principal clasts as lavas and pyroclastic rocks, together with various sedimentary lithologies, and concluded that they most resembled rocks from the Longmynd and other Shropshire, Welsh Borderland and Midlands Lower Palaeozoic and Precambrian outcrops. Lapworth (1898) noted that there was a significant amount of locally derived Silurian and

Devonian material in the Malvern–Abberley area. Ramsey (1855) noted possible Malvernian granites at Berrow Hill [744 585], north of their present outcrop. In outcrops south of the Malverns he also found Malvernian rocks, but in lesser abundance than the Longmyndian and other types. Fleet (1927) studied the mineralogy of various breccias and sandstones in the Midlands, including the Haffield Breccia. He concluded that they were 'definitely related to the Enville Breccias of Shropshire and Clent'. Worssam et al. (1989) noted that Malverns Complex lithologies were dominant in the south Malverns.

Ramsey (1855) suggested that the Permian breccias were of glacial origin on account of their poor stratification and sorting. Oldham (1894) and Lapworth (1898) proposed deposition of locally derived material by sheet floods and streams. The beds appear to comprise proximal, mixed debris-flow and fluvial sediments. Two sources have been proposed; Oldham and Lapworth suggested the Precambrian outcrops 45 to 50 km to the north-west; Wills (1956) suggested an easterly source, the Worcester Horst, which was uplifted and eroded, and now lies buried beneath the Worcester Basin. The proximal nature of the deposits is better explained by the latter model.

The age of these rocks is unproven. Ramsey (1855) and Lapworth (1898) assumed a Permian age, Fleet (1927), Wills (1948) and Mitchell et al. (1962) correlated them with the Upper Carboniferous 'Enville Beds'. Smith et al. (1974) and Worssam et al. (1989) tentatively assigned them to the Lower Permian. If the local, easterly source model is correct, and since they occur to the west of the East Malvern Fault, they must predate the rifting and formation of the Worcester Graben and thus may be no younger than Late Carboniferous in age.

There are few exposures in the outliers at Collins Green [740 573] and Berrow Hill [744 585]. A disused quarry [7403 5741] at Collins' Green exposes up to 5 m of ill-sorted, silty, poorly cemented breccia resting unconformably on Much Wenlock Limestone. Reeve (1953) reported clasts of Uriconian lithologies and possible Devonian cornstones. A temporary section [7430 5785] showed similar friable breccia faulted against the Bromsgrove Sandstone Formation. Field brash on the outliers comprises a wide range of igneous, volcanic, sandstone and grey quartzite clasts. Thin sections of some of the clasts in the BGS collections show them to be agglomerate (or conglomerate) consisting of fragments of andesites, oligoclase crystals, granite and dioritic rocks, tuff, mudstone and shale set in a ferruginous matrix with many quartz grains (E17624). The tuff clasts are variable; some contain fragments of felsitic and andesitic lava, some contain oligoclase (E17625, E17626) and some are banded tuff (E17627).

Bridgnorth Sandstone Formation

The Bridgnorth Sandstone (Warrington et al., 1980), formerly the Lower Mottled Sandstone, occurs at depth in the Worcester Basin, as proved by borehole and seismic data (Chadwick, 1985; Chadwick and Smith, 1988; Whittaker, 1980).

The Kempsey Borehole (Whittaker, 1980) proved 938.2 m of Bridgnorth Sandstone to a depth of 2294.8 m. Seismic evidence suggests a maximum of over 1000 m a little to the west, with thinning to the south-east (Figure 39). The gamma-ray and sonic logs from the borehole (Figure 21) have monotonous signatures (Penn, 1987), reflecting the uniform, well-sorted nature of the sandstone. The sonic log shows a greater range of values than the gamma-ray log, reflecting variations in the porosity.

The formation comprises pale reddish brown, very fine- to medium-grained, variably calcareous, dolomitic or siliceous sandstones. The grains are well rounded and consist dominantly of quartz and volcanic fragments. The dipmeter log of the Kempsey Borehole shows well developed large-scale cross stratification dipping at up to 30° and interpreted as aeolian in origin. Similar dune sets up to

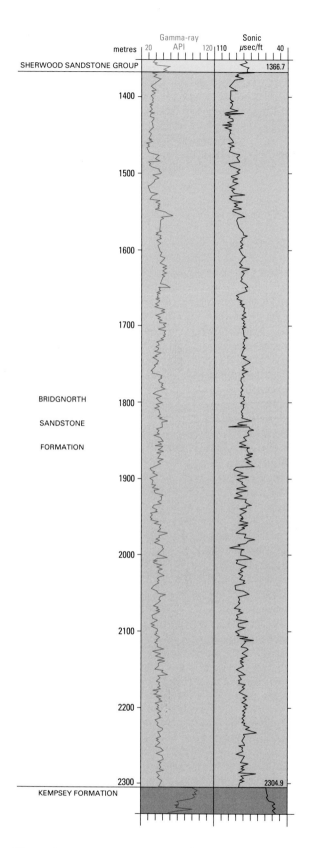

Figure 21 Wireline logs of the Bridgnorth Sandstone Formation, Kempsey Borehole.

4 m thick were observed at outcrop in the Tewkesbury area to the south (Worssam et al., 1989). The dip azimuths from the borehole show variable palaeowind directions. The dominant one is from the north-east, with some from the west, north-west and south-east. A similar dominant trend was recorded to the south (Worssam et al., 1989), and in the Kidderminster (Shotton, 1937; Mitchell et al., 1962) and Bridgnorth (Karpeta, 1990) districts to the north.

The formation is assumed to be Early Permian in age (Smith et al., 1974).

SHERWOOD SANDSTONE GROUP

The name was formally introduced by Warrington et al. (1980) for the arenaceous beds of the lower part of the British Triassic succession. They introduced three new formation names for the south Midlands area, which replace the classification of Hull (1869), shown in brackets. From base to top, these are the Kidderminster Formation (Bunter Pebble Beds), the Wildmoor Sandstone Formation (Upper Mottled Sandstone) and the Bromsgrove Sandstone Formation (Lower Keuper Sandstone). Only the Bromsgrove Sandstone has been recognised at outcrop, but all three formations have been proved in the Kempsey Borehole. Seismic reflection surveys show that they extend across the Worcester Basin within the district. The Wildmoor Sandstone is absent to the south (Ambrose, in preparation).

The gamma-ray and sonic logs of the Kempsey Borehole (Penn, 1987) are illustrated in Figure 22. Whittaker et al. (1985) subdivided the Sherwood Sandstone Group of the Wessex Basin (Units SS1–3) on the basis of its log signatures. However, these divisions do not correlate precisely with the lithostratigraphical succession at Kempsey. Unit SS1 equates with only the lower part of the Kidderminster Formation and Unit SS3 may be wholly or partially equivalent to the Sugarbrook Member of Old et al. (1991). The lowermost part of the Kidderminster Formation in the Kempsey Borehole, to 1288.6 m, has a distinctive gamma-ray signature with low values. Above this, the values increase and show a moderately serrated signature which continues up into the Wildmoor Sandstone and the lower part of the Bromsgrove Sandstone. A significant change occurs at about 692 m at the top of the Burcot Member, with a much more serrated signature above. This reflects the change from braided river to meandering channel deposits, with an increase in the number of argillaceous overbank sediments. Both upward-fining and upward-coarsening cycles can be identified on the gamma-ray log. In contrast, the sonic log has a very subdued and monotonous signature throughout, but becomes more serrated, with a greater range of velocities above 579 m, towards the top of the Bromsgrove Sandstone. There is overall a slight upward decrease in velocity.

There is a major unconformity within the Sherwood Sandstone Group to the south of the district, with the Wildmoor Sandstone and the Burcot Member of the Bromsgrove Sandstone Formation being cut out southwards (Ambrose, in preparation). Wills (1970, 1976)

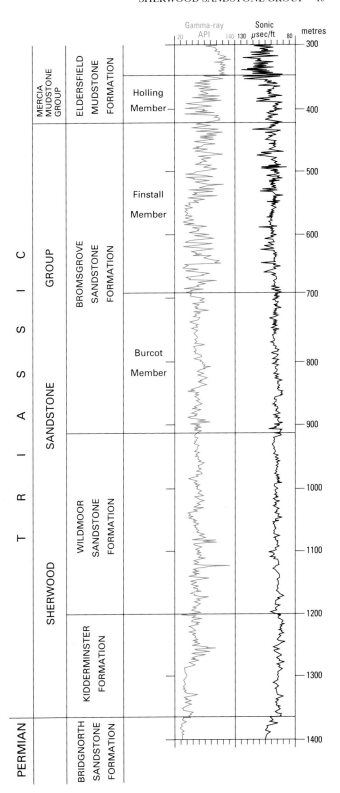

Figure 22 Wireline logs of the Sherwood Sandstone Group, Kempsey Borehole (modified after Penn, 1987).

identified an unconformity in the Stratford–Warwick and Birmingham areas and postulated a disconformity between the Bromsgrove and Wildmoor Sandstones in the Bromsgrove area (Old et al., 1991). The break, which cannot be identified in the Kempsey Borehole, has been equated with the Hardegsen Disconformity (Trusheim, 1963) in Germany (Warrington, 1970a).

No fossils have been found in the group within the district.

Kidderminster Formation

The Kidderminster Formation is known only from the cuttings samples of the Kempsey Borehole (Whittaker, 1980). They are of brown, medium- to very coarse-grained, poorly sorted, pebbly sandstones and conglomerates. There is some dolomitic cement, but the examples are mostly friable, with a clay matrix. The pebbles are well rounded and mostly of quartzite, with some volcanic material. Similar lithologies were recorded in outcrops to the north (Mitchell et al., 1962; Old et al., 1991). The thickness of the formation at Kempsey is 166.7 m. Isopachytes drawn using mainly seismic reflection data show that the thickest development of over 300 m is in the north of the district, immediately adjacent to the East Malvern Fault (Figure 39).

Wills (1970, 1976) subdivided the 'Bunter Pebble Beds' of Worcestershire into lower 'Shingle Beds' and upper, locally pebbly 'Middle Bunter'. These subdivisions may be represented by the twofold subdivision seen in the Kempsey Borehole gamma-ray logs (Figure 22). Old et al. (1991) noted upward-fining cycles with thin mudstone beds in the Bromsgrove–Lickey Hills area and concluded that the deposits are of fluvial, braided-river origin. Fitch et al. (1966) considered the river to have flowed northwards from the France–Bristol Channel area. Cross-bedding in the Kempsey Borehole indicates palaeocurrents mainly from the south and west. Whittaker (1980) suggested alluvial fan deposition, possibly with bursts of rapid run off. Fans may therefore have prograded east from the Malverns and Lower Palaeozoic outcrops, into a rift valley drained by a northward-flowing braided-river system. The presence of mud cracks and aeolian sand grains (Old et al., 1991) indicates an arid to semi-arid climate.

No fossils are known from the formation in the district. Trace fossils from the adjoining Droitwich district, considered to indicate a diverse land fauna including insects and reptiles, have been reviewed by King and Benton (1996).

Warrington et al. (1980) assigned an Early Triassic, Scythian, age to the Kidderminster Formation.

Wildmoor Sandstone Formation

The Wildmoor Sandstone Formation is known only from the Kempsey Borehole, where it comprises mainly reddish brown, very fine- to fine-grained, micaceous, well-sorted sandstones. They range from friable to moderately well cemented, with a dolomitic and patchily calcareous cement, and are coarser grained towards the base.

Interlaminations of siltstone and mudstone occur and there are rare intraformational conglomerates. Some greyish green sandstones are also present. The grains are well rounded and dominantly of quartz, with some volcaniclastic fragments. Sedimentary structures include cross-lamination and trough cross-bedding. Old et al. (1991) noted parallel lamination, rare red-brown mudstone beds and fining upward cycles at outcrop in the Bromsgrove–Lickey Hills area to the north.

The formation is 284.7 m thick at Kempsey, thickening north-westwards to over 600 m immediately adjacent to the East Malvern Fault (Figure 39). From seismic evidence, Worssam et al. (1989) suggested that the formation is about 300 m thick in the Tewkesbury district, but Ambrose (in preparation) demonstrated that the formation dies out south of the Worcester district, cut out by the Hardegsen Disconformity.

The sedimentary structures of the Wildmoor Sandstone suggest fluvial deposition, although the well rounded and even sized grains suggest reworking of aeolian deposits. Old et al. (1991) concluded that deposition occurred in a low-sinuosity fluvial environment, with seasonally active braided streams transporting material derived locally from sand dunes. Interpretation of the Kempsey dipmeter log suggests mainly braided streams with periodic meandering rivers. The dominant south and south-west dip direction in the borehole may be structural rather than depositional; around Bromsgrove, Hassan (1964) recorded a unimodal direction of transport from the south-east.

No fossils are known from the formation in the district. Fossils from the Droitwich district comprise undetermined plant stems from Whitford Mill Borehole [944 707] (F W Shotton, written communication to B C Worssam, 10th January 1977), and invertebrate trace fossils, vertebrate tracks and a fragmentary fish; see Pollard (1985), Old et al. (1991) and King and Benton (1996).

Warrington et al. (1980) assigned an Early Triassic, Scythian, age to the formation.

Bromsgrove Sandstone Formation

The Bromsgrove Sandstone crops out in seven areas adjacent to the East Malvern Fault. It has been proved at depth in the Kempsey Borehole and by seismic reflection surveys. Three of the outcrops probably represent the uppermost beds of the formation below the Mercia Mudstone Group. Others are entirely fault bounded and their age is unknown; some may belong to older formations of the Sherwood Sandstone Group. All of the outcrops are of a marginal facies and contain a greater proportion of conglomerates and breccias than the buried sediments in the centre of the basin.

In the Kempsey Borehole, the Bromsgrove Sandstone is 493.8 m thick. Isopachytes drawn mainly from seismic reflection profiles (Figure 39) show that the formation thickens eastwards from the East Malvern Fault, reaching a maximum close to the west of the Inkberrow Fault.

The formation was subdivided into three units: the Basement Beds, Building Stones and Waterstones (Hull, 1869; Warrington, 1968, 1970a). These units were formalised by

Old et al. (1991), in the Redditch district, as the Burcot, Finstall and Sugarbrook members respectively. They can be recognised in the Kempsey Borehole wireline logs (Figure 22), but the beds equivalent to the Sugarbrook Member are largely mudstone-dominated in the Worcester district, and are included in the Mercia Mudstone Group, where they are referred to the Holling Member (p.58).

Typical Bromsgrove Sandstone from the Kempsey Borehole comprises pale to medium reddish brown, mostly fine-grained, locally medium-grained, variably dolomitic and calcareous sandstone. The grains are moderately to well sorted and mostly of quartz, with minor volcanic fragments. The basal parts of some sandstones are medium to coarse grained. Thin interbeds of red siltstone and mudstone occur, particularly in the Finstall Member.

The heavy mineral suite of the Bromsgrove Sandstone contains relatively fresh minerals compared to older Triassic, Permian and Upper Carboniferous sandstones, and was, therefore, probably derived from a different source (Fleet, 1930).

The uppermost 50 m of the formation crop out in the north of the district, and comprise red to orange-brown, with subordinate grey-green, fine- to medium-grained, moderately sorted, friable sandstones, with subrounded to well-rounded grains. Some beds are sparsely pebbly, others contain conglomeratic layers or lenses. The pebbles are generally less than 10 mm in diameter and are mainly of igneous rocks, with some of quartz. Intraformational mudflake breccias also occur, along with some green and red, micaceous siltstones and silty mudstones. Most sandstones display large-scale tabular cross-bedding, with some parallel lamination.

South of Alfrick [749 531], up to 100 m of poorly sorted conglomerates and breccias are interbedded with sandstones and small amounts of mudstone and siltstone. Some of the sandstones are composed of well-rounded, medium-sized grains and are probably aeolian in origin. These beds are overlain, apparently conformably, by the Mercia Mudstone Group and are therefore assigned to the Bromsgrove Sandstone, but it is possible that they are overstepped by the Mercia Mudstone Group and may be older.

The fault-bounded outcrops comprise dominantly poorly sorted, clast-supported breccias and conglomerates in a sandy matrix. Some beds show a matrix-supported fabric. The clasts range from granules to boulders and from subangular to subrounded. Some beds are weakly cross-bedded but many are unbedded. Imbrication is generally poor to absent. Upward-fining cycles are common, with an upward passage from matrix-supported breccia/conglomerate to pebbly sandstone and sandstone. The latter are buff to pinkish grey, medium to fine grained, well sorted and cross-bedded, with subrounded grains. Channelised sandstones are common, with individual sand bodies ranging from 0.5 to over 4 m in thickness.

All of the breccias/conglomerates in these marginal deposits contain clasts derived from local Malverns Complex igneous rocks, volcanic rocks, Silurian sandstones and limestones. Fleet (1927) concluded that the breccias at Alfrick and Knightwick were related to the Enville Brec-

cias of Shropshire and Clent. Groom (1900), Wills (1948) and Phipps and Reeve (1967) correlated those at Knightwick with the Haffield Breccia, or Clent Group. There is a transition between these beds and the overlying sandstones at Knightwick, as recorded by Phipps and Reeve (1967). These authors and Groom (1900) referred the sandstones to the Bunter Sandstone (now Bridgnorth Sandstone), but they are more like the sandstones of the Bromsgrove Sandstone and are here assigned to that formation. On this basis, the breccias are therefore younger than the Haffield Breccia, and are included as a marginal facies of the Bromsgrove Sandstone.

Most of the Bromsgrove Sandstone consists of upward-fining fluvial cycles. Pebbly or conglomeratic bases represent channel lag deposits and the cross-bedded sandstones were deposited as migrating sand bodies within the channels. The finer-grained muddy sediments represent overbank deposits. In the Kempsey Borehole, the dipmeter log shows variable current azimuths in the lower part, up to 874.8 m, suggesting either alluvial fan or meandering channel facies. Above this, up to 691.9 m depth, the palaeocurrent directions are more unimodal and are interpreted as indicating deposition by a braided river, correlating with the Burcot Member of Old et al. (1991). The uppermost beds show more variable current directions, contain more argillaceous interbeds, similar to the Finstall Member of Old et al. (1991), and are interpreted as the deposits of meandering channel systems. The uppermost beds, cropping out north of the River Teme, were deposited at the basin margin in fining-upward fluvial cycles. To the south along the basin margin, this facies is interbedded with thicker conglomerates deposited in proximal fans.

In the most recent palaeogeographic reconstruction, Warrington and Ivimey-Cook (1992) envisage deposition in a major northward-flowing river, draining north-west France. In the Bromsgrove–Kidderminster area, palaeocurrent directions largely support this (Old et al., 1991). Variable flow directions were recorded in the upper part of the formation in the Bromsgrove area (Old et al., 1991, fig.7). In the Kempsey Borehole the dominant dip direction is to the south-west, but this may be structural rather than depositional. On the western basin margin the palaeocurrent directions are variable, but consistent with deposition in fans prograding eastwards. The main north-flowing river was probably confined tectonically to the central part of the graben, where the Bromsgrove Sandstone is thickest, between the Smite–Pirton–Tewkesbury fault system in the west and the Inkberrow Fault in the east (Whittaker, 1985; Chadwick and Smith, 1988).

No fossils were found in the district during the present survey. In its type area, around Bromsgrove, the formation contains equisetalean and coniferalean plants and a diverse microflora that reflects the presence of lycopsids, sphenopsids, pteropsids and gymnosperms. Microplankton were recovered from the Sugarbrook Member in the Sugarbrook No. 1 Borehole (p.59). The fauna comprises annelids, molluscs, arthropods, fish, amphibians and reptiles. A number of trace fossils has also been recorded. The composition and environmental significance of this biota have been reviewed by Old et al. (1991). In neighbouring districts, productive palynomorph samples were recovered from the Finstall Member between 23.47 and 81.23 m in the Westwood Borehole, from 17.68 m and 20.42 m in the Whitford Mill Borehole and at 376.65 to 376.68 m in the Eldersfield Borehole. These assemblages (Table 5) are comparable with those documented from outcrop at Hadley Mill [865 642], Elmley Lovett [870 696] and Rock Hill, Bromsgrove [948 698] (Clarke, 1965) and from the Sugarbrook No. 1 Borehole (Warrington 1970a; Old et al., 1991), Webheath No.1 Borehole (Old et al., 1991) and Knight's Lane Borehole (Warrington in Williams and Whittaker, 1974). The assemblages comprise miospores of land-plant origin. These include pteridophyte spores, but are dominated by bisaccate, mostly non-taeniate, pollen of gymnosperm origin, including morphotypes that have been isolated from male cones now assigned to the genus *Masculostrobus* (*M. bromsgrovensis*, *M. willsi*) (Wills, 1910; Townrow, 1962; Grauvogel-Stamm, 1972).

Warrington et al. (1980) assigned an Early to Mid-Triassic (Scythian to late Anisian/early Ladinian) age to the formation. Penn (1987) tentatively placed the Anisian–Ladinian boundary at the top of the formation in the Kempsey Borehole. Benton et al. (1994) placed the formation entirely within the Anisian, based on a re-view and reassessment of the vertebrate faunas and microfloras from the south Midlands. The occurrence of *Perotrilites minor*, *Angustisulcites* spp., and *Stellapollenites thiergartii* in miospore assemblages from the Westwood and Eldersfield boreholes (Table 5) is, by comparison with independently dated Triassic sequences elsewhere in Europe (Visscher and Brugman, 1981; Brugman, 1986), indicative of an Anisian (early Mid Triassic) age for the upper part of the Finstall Member and for the Sugarbrook Member. The age of beds in the lower part of the Finstall Member (below 81.23 m in the Westwood Borehole), and of the Burcot Member, is unproven.

The formation is generally poorly exposed throughout the district. Conglomerates and sandstones are well exposed around Knightwick immediately south [737 555] of the river Teme. The cliffs of Osebury Rock expose up to 35 m of conglomerates.

MERCIA MUDSTONE GROUP

The Mercia Mudstone Group crops out in a broad north–south belt in the centre of the district bounded in the west by the East Malvern Fault and isolated outcrops of the Bromsgrove Sandstone. In the east, younger rocks of the Penarth and Lias groups overlie the Mercia Mudstone. Topographically it is characterised by low scarps and gentle dip slopes. The Arden Sandstone Formation produces a more prominent scarp and the Blue Anchor Formation crops out on the scarp face of the most prominent escarpment, capped by the overlying Penarth Group and Wilmcote Limestone. The Severn and Teme valleys breach the outcrop which, in the west, is masked by Quaternary colluvial deposits extending up to 6km from the Malvern Hills.

The group consists predominantly of blocky red mudstones, silty mudstones and siltstones, with subordinate grey-green and laminated beds (Plates 8, 9). They weather to a red-brown, commonly heavy clay soil. Green reduction spots occur throughout and are common at some levels. Many have nuclei of radioactive minerals. Secondary satin spar gypsum veins are common throughout, and gypsum and anhydrite nodules are present at some levels. Borehole records show that gypsum is generally leached to a depth of about 30 m from the surface. Halite deposits, comprising the Droitwich Halite Member, have been proved immediately north of the district in the Saleway Borehole [9285 6017] (Wills, 1970, 1976; Poole and Williams, 1981) and are probably present at depth in the east of the district.

The group is estimated to be about 600 m thick over most of the district (Figure 36). Seismic data show that thicker accumulations are present in the east, where over 1100 m of beds are present, within which there is a significant unconformity (Chapter 10).

Murchison and Strickland (1840) first equated the 'Keuper Marl' and 'Keuper Sandstone' (now the Arden Sandstone) of Worcestershire and Warwickshire with the German Keuper. Hull (1869) outlined a Triassic stratigraphy that remained virtually unaltered until recently. Matley (1912) described the Arden Sandstone in detail around its type area to the north-east and Wills and Campbell-Smith (1913) described its flora and fauna. Matley followed Brodie (1870) in referring the beds above and below the Arden Sandstone respectively to the 'Upper' and 'Lower Marls'. Sherlock (1926) and Richardson (1929) used the term 'Upper Keuper Sandstone' for the Arden Sandstone and equivalents. Elliott (1961) subdivided the 'Keuper Series' of Nottinghamshire into eight formations, based largely on sedimentary structures. Some of these subdivisions were recognised in the Home Farm Borehole [SP4317 7309] in eastern Warwickshire (Old et al., 1987) and they can be tentatively recognised in the boreholes in the Worcester Basin sequence. Warrington (1967, 1970a) correlated the English Midlands sequences with those of the North Sea and north-west Europe on palynological evidence. Audley-Charles (1970a) subdivided the British Trias into five divisions and presented palaeogeographical reconstructions (1970b). More detailed palaeogeographical reconstructions and facies maps have been given by Warrington and Ivimey-Cook (1992). Wills (1970, 1976) subdivided the Midlands Triassic on the basis of cyclic sedimentation, recognising 15 'miocyclothems' in the 'Keuper Marl' of Worcestershire and Warwickshire. Warrington et al. (1980) revised the Triassic nomenclature of the British Isles, introducing the terms Mercia Mudstone Group, Arden Sandstone Member and Droitwich Halite Formation. Poole and Williams (1981) described the halite deposits at Droitwich. Details of the Mercia Mudstone Group to the south and north-east were given by Worssam et al. (1989) and Old et al. (1991) respectively. A revised lithostratigraphy for the group in the Worcester Basin and detailed correlation using wireline logs is given by (Ambrose, in preparation). New names are given to the red mudstones below and above the Arden Sandstone, the Eldersfield Mudstone and Twyning Mudstone formations respectively. The Arden

Table 5 Palynomorphs from the Bromsgrove Sandstone Formation (Sherwood Sandstone Group) and Eldersfield Mudstone Formation (Mercia Mudstone Group).

LOCALITY (SEE FOOTNOTE) / PREPARATION NUMBERS

Localities (top to bottom of table):

Locality	Preparation numbers
A	MPA 34209, 34211, 34214, 34230, 34231, 34233, 34236, 22790, 34238, 34241, 22791/34242, 34243, 34244, 22792/34245.6, 34247, 22793
B	MPA 21733
C	MPA 13429, 13432
D	SAL. 1541, 1094
E	SAL. 2438, 2439, 2440, 2441, 2442, 2443, 2444, 2445, 2446
F	MPA 21754, 21755

Miospores:

- *Alisporites* sp.
- *A. circulicorpus*
- *A. grauvogeli*
- *A. microreticulatus*
- *A. parvus*
- *A. toralis*
- *Angustisulcites* sp.
- *A. gorpii*
- *A. grandis*
- *A. klausii*
- *Apiculatasporites plicatus*
- *Aratrisporites* sp.
- *A. granulatus*
- *Calamospora* sp.
- *Colpectopollis ellipsoideus*
- *Convolutispora* sp.
- *Cyadopites* sp.
- *Cyclogranisporites* sp.
- *Cyclotriletes* sp.
- *C. microgranifer*
- *C. oligogranifer*
- *Densoisporites* sp.
- *D. nejburgii*
- *Illinites* sp.
- *I. chitonoides*
- *I. kosankei*
- *Kraeuselisporites* sp.
- *K. hoofddijkensis*
- *Labiisporites granulatus*
- *Lunatisporites* sp.
- *L. pellucidus*
- *Ovalipollis pseudoalatus*
- *Perotrilites minor*
- *Porcellispora longdonensis*
- *Protodiploxypinus* sp.
- *P. doubingeri*
- *P. fastidiosus*
- *P. potoniei*
- *P. sittleri*
- *Retisulcites perforatus*
- *Staurosaccites* sp.
- *S. quadrifidus*
- *Stellapollenites thiergartii*
- *Striatoabieites* sp.
- *S. balmei*
- *Sulcatisporites* sp.
- *S. kraeuseli*
- *Triadispora* sp.
- *T. crassa*
- *T. epigona*
- *T. falcata*
- *T. plicata*
- *Tsugaepollenites oriens*
- *Verrucosisporites* sp.
- *V. contactus*
- *V. jenensis*
- *V. krempii*
- *V. remyanus*
- *Voltziaceaesporites heteromorpha*
- indeterminate trilete spores
- indeterminate bisaccate pollen

Organic-walled microplankton and algae:

- *Cymatiosphaera* sp.
- *Dictyotidium* sp.
- *Tasmanites* sp.
- *Veryhachium* sp.
- *Plaesiodictyon mosellanum*

DEPTHS (M) (BOREHOLE SAMPLES):

Preparation	Depth (m)
22793	300.6 – 300.9
34247	293.3 – 293.5
22792/34245.6	289.25 – 290.0
34244	287.2 – 287.3
34243	275.3 – 275.5
22791/34242	257.35 – 257.75
34241	248.82 – 249.41
34238	229.03 – 229.45
22790	224.8 – 225.0
34236	217.18 – 217.35
34233	205.9 – 206.0
34231	203.00 – 203.08
34230	202.00 – 202.25
34214	132.70 – 132.75
34211	118.25 – 120.00
34209	110.36 – 110.56
MPA 13429/13432	351.20 / 376.20 – 376.68
SAL. 1541/1094	57.00 / 57.61
SAL. 2438	23.47
2439	27.43
2440	27.89
2441	36.58
2442	60.43
2443	64.29
2444	71.02
2445	78.03
2446	81.23
MPA 21754	17.68
21755	20.42

LITHOSTRATIGRAPHY:

- ELDERSFIELD MUDSTONE FORMATION — MERCIA MUDSTONE GROUP
- BROMSGROVE SANDSTONE FORMATION — SHERWOOD SANDSTONE GROUP

Footnotes:

A Worcester Heat Flow Borehole
B Hartlebury brickpit
C Eldersfield Borehole
D Webheath No. 1 Borehole
E Westwood Borehole
F Whitford Mill Borehole

? uncertain record
cf comparable form

Plate 8   Belmont Brick Pit [771 476], Great Malvern, in 1950, showing minor faulting and folding in the Eldersfield Mudstone Formation (Mercia Mudstone Group). The pit is now backfilled with domestic refuse and built on. A 8464

Sandstone is elevated to formation status and the Droitwich Halite is relegated to member status. The beds formerly termed the 'Waterstones' are named the Holling Member and form the lowermost part of the Eldersfield Mudstone Formation.

In the Worcester Basin, the group consists of alternations of relatively hard and soft red mudstones, the former giving rise to topographic ridges and dip slopes. Green mudstones, siltstones and sandstones are relatively uncommon. Subdivision of the group has been made from a cored borehole at Worcester [8624 5762], with additional information from other cored sequences to the south at Twyning [8943 3664] and Eldersfield [7891 3221].

MINERALOGY

The results of a limited amount of X-ray, sedigraph and particle size analysis of samples of red mudstone from the Eldersfield Mudstone and Twyning Mudstone formations suggest that the harder, feature-forming mudstones are slightly coarser grained than the intervening softer ones. Variations in the amount and nature of the iron cement may also influence hardness and resistance to weathering. Both mudstone types have comparable dolomite contents and this is therefore unlikely to have any effect on the weathering properties.

The bulk of the grains are silt-size, with some coarser grains, up to coarse sand-size. They vary from subrounded to well rounded and are predominantly of quartz (over 80%). Microprobe analysis has shown that feldspars with compositions of near pure albite and orthoclase dominate the non-quartz fraction. Other grains include muscovite, ilmenite, magnetite, dolomite and calcite. Calcite usually occurs as crystal aggregates. Much of the quartz is contaminated with hematite and clay minerals. The X-rays show abundant quartz and dolomite, with calcite, illite and chlorite. Dumbleton and West (1966) and Jeans (1978) studied the clay mineralogy of the Mercia Mudstone Group, including samples from a number of sites in Worcestershire. They concluded that the matrix or primary mineral was illite, possibly with small amounts of

Plate 9   Entrance to Colwall railway tunnel [774 436], Upper Wyche, in 1925, showing minor folding and faulting in the Eldersfield Mudstone Formation (Mercia Mudstone Group). A 3243

chlorite. These were present, together with calcite and/or dolomite and quartz, in all samples analysed. Approximate bulk compositions (Dumbleton and West, 1966) are: illite — 30–60%; chlorite and swelling chlorite — up to 40%; dolomite — up to 25%; calcite — up to 17%; quartz — 6-35%; hematite — up to 2% (not present in green beds). Dumbleton and West (1966) suggested that the chlorites are secondary, resulting from the transformation of illite by high magnesium concentrations. Exotic clay minerals include sepiolite, smectite, corrensite and palygorskite. Jeans (1978) suggested that the variations in the exotic assemblages may be related to salinity gradients within the basin.

SEDIMENTOLOGY

The Eldersfield Mudstone and Twyning Mudstone formations consist mainly of alternating blocky and laminated mudstones and siltstones. These facies were described by Arthurton (1980) in the Mercia Mudstone Group of the Cheshire Basin. In addition, laminated

sandstones, with minor siltstone and mudstone, are present but are largely confined to the Holling Member at the base of the Eldersfield Mudstone Formation.

The blocky facies comprises structureless mudstones and siltstones, in which both upward-fining and upward-coarsening sequences occur. Individual units range between 1 and 9 m in thickness. Upward-coarsening units are more common, and are usually truncated by erosion surfaces overlain by laminated beds. The bases of blocky units contain reworked clasts of underlying laminites. The clasts are angular, matrix-supported and decrease in size and abundance upwards; they are the product of desiccation and redistribution by wind. Desiccation cracks are rarely evident in the blocky beds, but some penetrate down from the overlying laminites. Green reduction spots are locally common and well-rounded sand grains occur locally. The blocky beds are interpreted as aeolian, grain size variation perhaps representing variation in wind strength and direction. Impersistent, weak lamination is seen locally, with preferential dolomitisation of the coarser laminae. These beds may represent subaerial deposition of adhesion rippled sediment, or localised lacustrine deposition in shallow ponds, or small-scale sheetflood deposition.

The blocky facies with weak lamination is common in the lower part of the Worcester Borehole, below 158 m, and is, in places, disrupted by soft-sediment deformation which may have resulted from seismic activity or rapid dewatering, the latter perhaps triggered by the former.

The laminated facies (Plates 10b, d–h, 11) is mostly colour-laminated, red-brown and mid to pale greenish grey, and, locally, brick red, pinkish red and purple-brown. The coarser laminae are commonly dolomitic and grade upwards from fine-grained sandstone to siltstone. The beds are mostly parallel laminated, but ripple lamination, small-scale cut and fill, and load structures and other thixotropic features also occur, as well as pseudomorphs after halite. Desiccation cracks are common.

Below 158.44 m in the Eldersfield Mudstone of the Worcester Borehole many of the laminated beds are highly disturbed or brecciated (Plate 10g, h). These disrupted beds resemble those in Nottinghamshire described by Elliott (1961) as penecontemporaneous 'flow breccias'. Wilson (1990) described similar breccias in the Mercia Mudstone Group of the Eastern Irish Sea Basin and interpreted them as the product of dissolution of halite. The Worcester examples may also be dissolution breccias, although some of them may have been caused by seismic activity associated with syndepositional faulting.

The sandstones in the sandstone-mudstone laminites (Plate 11a, b, c) at the base of the Eldersfield Mudstone Formation are generally parallel-laminated and locally micaceous, with some planar cross-laminated beds, trough cross-laminated beds and massive beds (Plate 11b). Burrowing is generally restricted to these basal beds.

The laminated beds are interpreted as playa lake deposits, subjected to frequent subaerial exposure and desiccation.

A range of pedogenic and evaporitic features is present in the Eldersfield Mudstone Formation of the Worcester Borehole. Gypsum and anhydrite nodules occur throughout. They are rare in the laminated facies in the Worcester Borehole, but more common in the Eldersfield Borehole. Wright et al. (1988) interpreted these nodules in the Mercia Mudstone Group of southwest England as having formed in palaeosols in a playa-alluvial setting, although Leslie et al. (1993) considered it unlikely that pedogenic processes were active throughout deposition of the upper part of the Mercia Mudstone Group. Brecciated horizons comprising paler dolomitic mudstone with a dense ramifying network of fine, dark red, poorly dolomitic mudstone veins (Plate 10a) are common at some levels. These appear to be desiccation breccias, probably formed by repeated shrinkage and expansion caused by wetting and drying of lithified sediment. Examples of vein-like networks fanning down from the base of desiccation cracks also suggest an origin by desiccation. Elliott (1961) described similar breccias in Nottinghamshire which he termed 'contemporaneous vein-type breccias'.

Calcite nodules occur in the lowermost beds of the Worcester Borehole, below 266.15 m. From 266.15 m to 287.60 m they occur with anhydrite, below which they occur alone. Dolomite is the dominant carbonate above 266.15 m, occurring as discrete nodules and diffuse patches, some of which have a brecciated texture. This carbonate probably formed as pedogenic calcrete, with subsequent wetting and drying causing in-situ brecciation.

PALAEONTOLOGY

The principal fossils recovered from the Group in the district are palynomorphs from the Eldersfield Mudstone Formation and palynomorphs and trace fossils (burrows and footprints) from the Arden Sandstone Formation. Material that aids interpretation of the Worcester sequences has also been recovered from the Eldersfield Mudstone Formation in the Droitwich district, from the Arden Sandstone Formation in the Tewkesbury and Redditch districts and from the Blue Anchor Formation in the Droitwich and Redditch districts. Macrofossils are very sparse, but include plant and vertebrate remains from the Holling Member and the Arden Sandstone and Blue Anchor formations. The palynomorphs indicate an age range of Anisian to Rhaetian.

CONDITIONS OF DEPOSITION

The depositional environments for the Mercia Mudstone Group are summarised in Table 6. The red-bed sequences were mostly deposited in sabkha to playa environments in an arid to semi-arid climate. Acritarchs and an alga in the Holling Member and immediately overlying beds (Table 5; p.59) indicate marine influence and brackish-water conditions during initial deposition. The green beds of the Arden Sandstone and Blue Anchor formations probably represent pluvial conditions and a more humid climate (Simms and Ruffell, 1990).

The halite deposits are interpreted as forming in very shallow marine waters, either during a period of raised sea level that also resulted in halite deposition in Cheshire, the Irish Sea, Staffordshire and Somerset, or during rifting in sub-sea level grabens.

Plate 10 Lithologies of the Eldersfield Mudstone Formation, Worcester Borehole.

a. Blocky, dolomitic mudstone with brecciated texture and reduction spots.
b. Laminated mudstone with desiccation cracks overlain by blocky mudstone with fragments of reworked laminated mudstone.
c. Blocky, weakly dolomitic mudstone with nodular anhydrite and green reduction spots. There are some vague bedding traces and faint desiccation cracks at the base.
d. Laminated beds with desiccation cracks.
e. Finely laminated sandstone, siltstone and mudstone from the lower part of the Eldersfield Mudstone Formation.
f. Ripple laminated sandstone cut by a fibrous gypsum vein.
g, h Disrupted laminite with fibrous gypsum veins and anhydrite nodules.

Table 6 Depositional environments of the Mercia Mudstone Group. The main environment is shown first.

Blue Anchor Formation	Lacustrine
Twyning Mudstone Formation	Sabkha–Playa Lake
Arden Sandstone Formation	Fluvial–estuarine
Eldersfield Mudstone Formation Divisions 3–6	Sabkha–Playa Lake
Division 2	?lacustrine–estuarine/ intertidal
Holling Member (Division 1)	Estuarine/intertidal–Sabkha

GEOPHYSICAL CORRELATIONS

The geophysical logs (Figure 23) do not provide definite correlations between the boreholes shown, or with those from farther afield, due to the poor quality of some of the data, the absence of sonic logs of the Eldersfield Mudstone Formation, and probable lateral variation in this formation. There are, however, some reliable geophysical markers in the Eldersfield Mudstone Formation (Figure 23c, d, f) and some (Figure 23r, s, t, u, v) in the Twyning Mudstone Formation.

Lott et al. (1982) subdivided the Mercia Mudstone Group of the western Wessex Basin into six lithostratigraphical divisions (Units A–F), largely on geophysical characters and cuttings samples. Whittaker et al. (1985) recognised these units in the southern part of the Worcester Basin. The geophysical markers on which these subdivisions are based can be traced into the centre of the basin, and others allow further refinement (Ambrose, in preparation). The gamma-ray logs of Worcester, Twyning and Eldersfield and the sonic log of Worcester have been filtered to remove excess background 'noise' and a casing correction has been applied to the gamma-ray logs of Kempsey and Eldersfield.

STRATIGRAPHY

Eldersfield Mudstone Formation

The Eldersfield Mudstone Formation is named from the BGS Eldersfield Borehole [7891 3221]. Its base is that defined by Warrington et al. (1980) for the Mercia Mudstone Group, where mudstone and siltstone become dominant over sandstone; in the borehole this lies at 368.53 m depth.

The formation comprises dominantly red mudstones. It includes, at its base, the Holling Member (formerly 'Waterstones'). This member correlates with the Sugarbrook Member of the Redditch district, which is sandstone dominated and included in the Bromsgrove Sandstone. It crops out in the north of the district near Martley, and is named from Upper Holling Farm, where a borehole [7537 5828] provides the type section. In the

Kempsey Borehole, sandstone forms an estimated 50 per cent of the member, and Whittaker (1980) included it in the Bromsgrove Sandstone. Penn (1987) included it in the Mercia Mudstone and referred it to the Tarporley Siltstone of the Cheshire Basin. In the Eldersfield Borehole, the member contains only 26 per cent sandstone, but was included in the Bromsgrove Sandstone by Worssam et al. (1989).

The formation makes up most of the Mercia Mudstone Group outcrop, with the oldest beds cropping out in the north-west. In the east the outcrop is disrupted by faulting, with up-faulted blocks within the overlying Arden Sandstone and Twyning formations, and inliers in the Brook End–Hatfield area and in the south around Upton on Severn.

The topography is dominated by scarps and east to south-easterly dip slopes formed by more resistant beds. These are best developed in the lowermost beds of the formation in the extreme north-west and, to a lesser extent, in higher beds in and around Worcester. More or less flat-lying beds occur in the Teme valley near its confluence with the Severn and in the city of Worcester.

Full thicknesses of the formation are known from two boreholes within the district: 401.96 m in the Upton on Severn Borehole [8341 4043] (Richardson, 1923) and about 371 m at Kempsey. To the south of the area, the Eldersfield Borehole proved about 366 m. Seismic reflection profiles show a considerable eastward thickening in the northern part of the district and an unconformity in the upper part of the formation (Figure 40). The Holling Member ranges from 10.9 m in the Upton on Severn Borehole to 73.15 m in the Kempsey Borehole (Whittaker, 1980), the latter thickness being based on geophysical and cuttings data.

The formation is poorly fossiliferous and within the district has only yielded palynomorph assemblages, fish debris and the crustacean *Euestheria minuta* from the Holling Member in the Worcester Borehole. Beds low in the formation are exposed in a brickpit [851 709] at Hartlebury, some 12 km to the north, in the Droitwich district, where palynomorphs were recovered (Table 5) 18 m above the base of the section. The Worcester Borehole proved a 300 m sequence terminating in the highest beds of the Holling Member. Palynomorphs have been recovered from the lower 190 m, principally between 110.36 and 120.00 m, 202.00 and 229.45 m and 257.35 and 290 m (Table 5). Miospore associations below 200 m, and those from the Hartlebury section, are similar to those from the underlying Bromsgrove Sandstone Formation. The occurrence of *Perotrilites minor* and *Stellapollenites thiergartii* with *Angustisulcites* spp. below 202 m in the Worcester Borehole, and of *S. thiergartii* with *Angustisulcites klausii* at Hartlebury indicates, by comparison with records from elsewhere in Europe (Visscher and Brugman, 1981; Brugman, 1986), that the lower part of the formation, like the upper part of the Bromsgrove Sandstone Formation, is of Anisian (early Mid Triassic) age. Acritarchs (*Cymatiosphaera, Dictyotidium, Veryhachium*) and tasmanitid algae have been recovered from 287.2 m, 290.0 m and 293.3 to 293.5 m. *Plaesiodictyon mosellanum*, a possible green alga of the

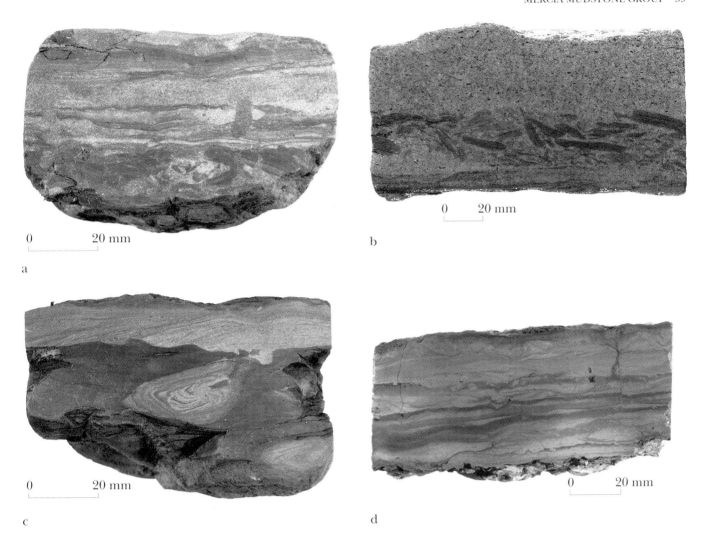

a

b

c

d

Plate 11 Lithologies of the Eldersfield Mudstone Formation, Worcester Borehole.

a. Sandstone with mudstone laminae and burrow traces from close above the Holling Member.
b. Apparantly massive sandstone with basal mud flake breccia, Holling Member.
c. Cross-laminated sandstone with loaded base overlying silty mudstone with detached sandstone
   load ball, Holling Member.
d. Fine-grained, ripple laminated and convoluted sandstone, interlaminated with siltstone and
   silty mudstone. Pseudomorphs after halite occur in the mudstone laminae.

Order Chlorococcales (Wille, 1970) occurs between 257.35 and 300.9 m. *Veryhachium* is represented by a solitary dark specimen that is probably reworked from Lower Palaeozoic deposits, but *Cymatiosphaera* and *Dictyotidium* are considered indigenous to the Eldersfield Mudstone Formation. These remains indicate a marine influence during deposition of the Holling Member and immediately overlying beds. Similar remains recovered from the Sugarbrook Member in the Redditch district (Warrington, 1967; 1970a; Old et al., 1991) indicate persistence of that influence across the Bromsgrove Sandstone Formation–Eldersfield Mudstone Formation (Sherwood Sandstone Group–Mercia Mudstone Group) interface. *Plaesiodictyon mosellanum* indicates brackish-water conditions.

The sequence above 202 m in the Worcester Borehole has yielded few miospore assemblages. The occurrence of *Retisulcites perforatus* at 110.36 to 110.56 m indicates a Ladinian (late Mid Triassic) to earliest Carnian (early Late Triassic) age. This miospore occurs just above geophysical marker 'i' (Figure 23), which is correlated with the Cotgrave Member of the East Midlands (Ambrose, in preparation), and above which the miospore also occurs. Palynomorph assemblages have been recovered from higher beds in the Saleway Borehole (Table 7). Assemblages at 218.85 m, 227.69 m and 231.34 m are comparable with ones recovered elsewhere from the Arden Sandstone Formation, the exact position of which in this borehole is uncertain (p.64). The presence of *Duplicisporites* spp., *Camerosporites secatus* and *Vallasporites ignacii*,

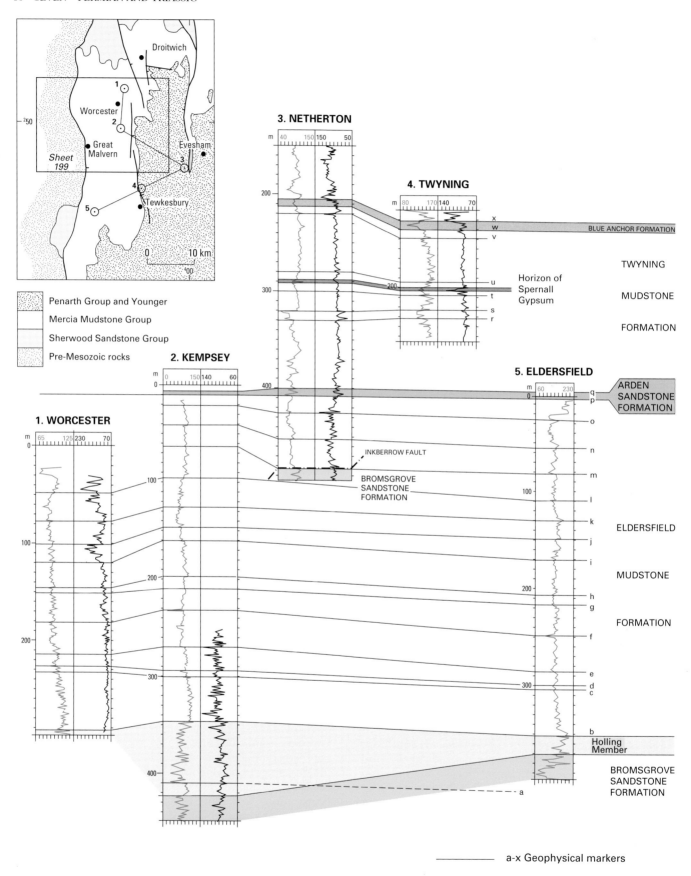

Figure 23 Correlation of the Mercia Mudstone Group using gamma-ray (in sepia; APi units) and sonic (in black; μsec/ft) logs of boreholes in the Worcester and adjacent districts.

LOCALITY (SEE FOOTNOTE)	A	B			C					D						E				
PREPARATION (MPA) NUMBERS	22911	25631	25632	25633	20411	20412	20413	20414	20415	25609	25610	25611	25612	25613	25614	19831	19832	19833	19834		
Miospores															(BARREN)	(BARREN)					
Porcellispora longdonensis	+	+	+	+	+	+	+	+				?					+	+	+		
Duplicisporites granulatus	+	+	+	+	+	?	+	+									+	+	+	+	
D. verrucosus	+	+	+	+	+			+									+	+	?	?	
Haberkornia gudati	?	+	+	+	+			+									+	+	+	?	
Camerosporites secatus	+	+	+	+	+			+									+	+	+		
indeterminate circumpolles	+																				
Vallasporites ignacii	+	+	+	+	+	?	+	+				+					+	+	+	+	
Ovalipollis pseudoalatus	+	+	+	+	+	+	+	+									+	+	+	+	
Ellipsovelatisporites plicatus	+	+	+	+	+	+											+	+	+	+	
Triadispora barbata	?																				
Rimaesporites potoniei	?	?	?		+	+						+					+	+			
indeterminate non-taeniate bisaccates	+	+	+	+	+	+					+	+					+				
Ricciisporites umbonatus	+	+	+	+								+									
Brodispora striata	?	+	+	+	+																
Calamospora mesozoica		+	+																		
Verrucosisporites contactus		+	+																		
indeterminate trilete spores		+			+																
Aratrisporites fimbriatus		+																			
A. saturni		+																			
Duplicisporites scurrilis		+	+	+																	
Patinasporites densus		+	+	+	+	+	+														
Protodiploxypinus sp.		? cf	+	+	cf +	+															
Angustisulcites sp.		? cf	+	+	+																
Cuneatisporites radialis		+	+		+							+					?	+			
Parvisaccites triassicus					+	+						+					+	+			
Protohaploxypinus sp.		+																			
Enzonalasporites vigens					?												+				
Aratrisporites sp.					? ?												?				
Alisporites toralis					+ +												? +				
Klausipollenites devolvens					+ +												+	+			
Enzonalasporites sp.												?						?			
Calamospora sp.																	+	+			
Deltoidospora sp.																	+	?			
Partitisporites quadruplicis																	?	+			
Patinasporites sp.																	+	?			
Triadispora sp.																	+	+			
T. plicata																	+				
Alisporites sp.																	?				
Angustisulcites klausii																	?				
Ellipsovelatisporites sp.																	?				
Retusotriletes sp.																					
Verrucosisporites morulae																	+				
Osmundacidites alpinus																	+				
Praecirculina granifer																					
Haberkornia parva																	cf				
Podosporites amicus																	+				
Ellipsovelatisporites rugosus																	?				
Triadispora aurea																	+				
T. obscura																	+				
T. stabilis																	+				
T. vilis																	+				
Klausipollenites sp.																	cf cf				
Sulcatisporites sp.																	? +				
Lunatisporites acutus									+												
Striatoabieites balmei									+												
Lunatisporites sp.																	? +				
Organic-walled microplankton and algae																					
Cymatiosphaera sp.	?																				
Botryococcus		+	+		+													+	+		
Plaesiodictyon mosellanum	+	+	+		+								+	+			+	+	+		
DEPTHS (M) (BOREHOLE SAMPLES)					311	312	313	314	314.5–314.9							214.88	218.85	227.69	231.34		
LITHOSTRATIGRAPHY		Arden Sandstone Formation														EMF					

A — The Grove, Upton upon Severn   D — Offerton Farm
B — Blue Hill railway cutting   E — Saleway Borehole
C — Twyning borehole   For borehole locations see Figure 20

? — uncertain record
cf — comparable form

EMF   Eldersfield Mudstone Formation

Table 7   Palynomorphs from the Eldersfield Mudstone Formation (above the Droitwich Halite) and the Arden Sandstone Formation.

in association with *Ovalipollis pseudoalatus* and *Ellipsovelatisporites plicatus*, indicates a Carnian (early Late Triassic) age.

The palynomorph assemblages recovered from the Bromsgrove Sandstone and Eldersfield Mudstone formations in the Worcester and neighbouring districts allow correlation with Triassic successions to the north-west that aid interpretation of the age of the Worcester sequence. In west Lancashire *Perotrilites minor* occurs, with *Stellapollenites thiergartii*, *Angustisulcites* spp. and *Tsugaepollenites oriens*, in the lower part of the Kirkham Mudstone Formation (Mercia Mudstone Group), below the Preesall Salt (Warrington, 1974; Warrington in Wilson and Evans, 1990). In the upper part of that formation, above the Preesall Salt, *Angustisulcites* spp., *S. thiergartii* and *T. oriens* persist and *Retisulcites perforatus* appears. Thus, the occurrence of *R. perforatus* at 202 m in the Worcester Borehole indicates that correlatives of the Preesall Salt of west Lancashire, and the Northwich Halite of Cheshire, occur below that level and that higher beds, including the Droitwich Halite, correlate with younger beds in those sequences (p.64). The presence of *T. oriens* in the upper beds of the Bromsgrove Sandstone Formation in the Webheath No. 1 Borehole (Table 5) and the Knight's Lane Borehole (Warrington in Williams and Whittaker, 1974) indicates a correlation with beds at or above the base of the Kirkham Mudstone Formation in west Lancashire, thus confirming the general southward younging of the Sherwood Sandstone Group–Mercia Mudstone Group interface proposed by Warrington (1970b).

In the Worcester Borehole, the Eldersfield Mudstone Formation can be subdivided into six divisions on lithological and pedogenic criteria (Figure 24).

Division 1   301.32 to 293.51 m

This is the topmost part of the Holling Member. It consists of interbedded fine-grained sandstones, siltstones and mudstones, with sandstones forming about 50 per cent of the member. Upward-fining sandstone/mudstone cycles represent a transition from shallow-water, traction-dominated, fluvial or lacustrine deposition to aeolian deposition, with subaerial exposure leading to early lithification and desiccation, perhaps in a coastal plain setting. The presence of marine microplankton (acritarchs) suggests shallow marine or brackish conditions (Old et al., 1991; Warrington and Ivimey-Cook, 1992).

Division 2   293.51 to 285.35 m

This division can be correlated on lithological grounds with the Radcliffe Formation (Elliott, 1961) of the East Midlands. It consists mainly of laminated beds interbedded with sandy laminites and, more rarely, blocky beds. This suggests a dominantly lacustrine environment, subject to frequent subaerial exposure and desiccation, with few subaerially accreted beds. A marine influence is indicated by the presence of acritarchs. Calcite nodules occur below 287.60 m, gypsum and anhydrite nodules occur above, with some traces of calcite. Dolomitic concretions occur at two levels.

Division 3   285.35 to 199.52 m

In this and succeeding subdivisions, blocky beds predominate. This unit is mostly dolomitic and contains disrupted beds (Plate 10g, h) and undisrupted laminites. The blocky facies occurs in upward-coarsening sequences in which dolomitic concretions are common. In-situ brecciated beds occur at two levels, anhydrite nodules are common throughout and mostly only a few millimetres in diameter. Below 266.15 m, some of the nodules are slightly calcareous.

Division 4   199.52 to 159.46 m

This unit contains a few laminated beds and beds with relict lamination, but both become increasingly disrupted downwards. Upward-coarsening sequences are common. Pedogenically brecciated horizons and dolomitic beds are present, the former becoming commoner upwards, the latter less so. Gypsum and anhydrite nodules are absent. The top is gradational, with a gradual increase in laminated beds towards the top, and also a decrease in the abundance of dolomite concretions (above 157.8 m) and in the abundance of laminated beds (above 158.44 m).

The disrupted beds of Division 3 and the lower part of Division 4 resemble those described by Elliott (1961) as penecontemporaneous flow-type breccias in the Carlton Formation of the Nottingham district, which lies at a similar stratigraphic level.

Division 5   159.46 to 75.99 m

This is characterised by common laminated beds and desiccation breccias, with some gypsum and anhydrite nodules and dolomite concretions. The top is gradational as the laminated beds gradually decrease in abundance upwards. The unit may equate with the Harlequin Formation of Elliott (1961), although breccias were not recorded in Nottinghamshire.

Division 6   75.99 to 30.00 m

This unit is predominantly blocky, with only sporadic laminated beds, many of which are fragmented and represented by mudstone clasts. Laminated beds are, however, relatively common between 57.87 m and 48.60 m. There are only a few breccia horizons, and gypsum and anhydrite nodules are also rare. The top of this division was not proved, coring of the borehole starting at 30 m depth. In the Eldersfield Borehole, similar beds continue to the top of the Eldersfield Mudstone Formation.

Divisions 3 to 6 contain typical sabkha–playa lake facies associations of alternating blocky and laminated

Figure 24 Correlation of the Eldersfield Mudstone Formation of the Worcester and Eldersfield boreholes, showing lithological variations and their relation to the wireline logs. The sonic log of the Eldersfield Borehole covers only the lowermost section (below 297 m) and is not illustrated. Dolomitic patches and concretions are likely to be present in the Eldersfield Borehole, but are not recorded.

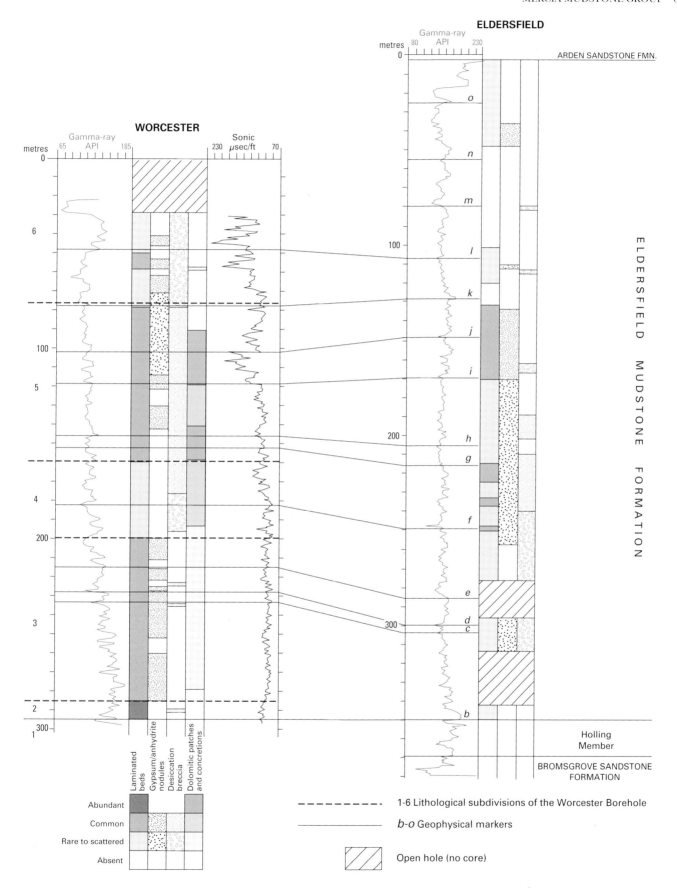

facies. The laminated beds commonly have desiccated tops, overlain by blocky mudstones with reworked mudstone flakes in their basal parts.

The lithological subdivisions of the Worcester Borehole are less clear in the Eldersfield Borehole (Figure 24), apart from the Holling Member (Division 1), which is represented by 20.06 m of mudstone, siltstone and sandstone between 368.53 m and 348.47 m depth. Divisions 2 and 3 are not easily distinguished. This is partly due to the absence of core, but the well-laminated Division 2 is poorly developed and the abundance of small anhydrite nodules and disrupted bedding characteristic of Division 3 is not seen. The dominantly blocky Division 4 is represented by the beds between 270.2 m and 222.9 m, which contain a few gypsum/anhydrite nodules and some intervals in which laminated beds are common, but it is lithologically indistinguishable from any of the underlying beds cored down to the top of the Holling Member. The sequence from 229.9 m up to 129.1 m is correlated with Division 5, although laminated beds are common only in the upper part and there are few brecciated horizons. The beds from 129.1 m in the Eldersfield Borehole up to the Arden Sandstone Formation are predominantly blocky and are correlated with Division 6.

Regional correlation of geophysical logs suggests that the Worcester Borehole commenced in beds equivalent to those near the top of the halite beds of the Wessex Basin, which are correlated with the Droitwich Halite Member (Ambrose, in preparation). Marker 'i' is at about 119 m in the Worcester Borehole and about 170 m in the Eldersfield Borehole (Figures 23, 24). This marker lies approximately at the base of the halite beds of the Somerset Basin. There is a marked increase in the abundance of gypsum/anhydrite nodules above this level at Eldersfield, but not at Worcester.

There are generally fewer gypsum/anhydrite nodules, laminated beds and desiccation breccias in the Eldersfield Borehole, compared to the Worcester Borehole, and no dolomite concretions were recorded in the former. The decrease in abundance of laminated and desiccated beds continues southwards to the Stowell Park Borehole near Cirencester and eastwards to the Knight's Lane Borehole near Stratford upon Avon (Williams and Whittaker, 1974). Laminated beds are common in the Putcheon's Farm Borehole near Redditch (Poole, 1969).

The Droitwich Halite Member has not been proved in the Worcester district, but to the north it is 158.57 m thick in the nearby Saleway Borehole (Wills, 1970; 1976; Poole and Williams, 1981), and evidence of salt subsidence was noted near the northern margin of the district [889 595], immediately east of the Smite Fault (Figure 32). The absence of subsidence features west of the fault suggests that it was an active growth fault during Triassic times and controlled deposition of the halite. The Inkberrow Fault is thought to have acted as the eastern boundary to halite deposition. The southward extent of the halite is not known, but the Mercia Mudstone Group is up to 1200 m thick east of the Smite–Pirton–Tewkesbury Fault system and likely to

contain halite. The BGS Stratford seismic line shows some strong reflectors within the Eldersfield Mudstone Formation which may be halite beds (Chapter 10, Figure 40).

At Saleway, halite occurs within the Droitwich Halite Member as discrete beds up to 11.86 m thick and as veins and nodules in mostly blocky mudstones; the thicker beds contain mudstone inclusions.

There is no direct evidence of the age of the Droitwich Halite, but it is assumed to be latest Ladinian or possibly Carnian on the basis of ages assigned to palynomorphs recovered from the Eldersfield Mudstone in the Worcester Borehole and from the upper 20 m of that formation in the Saleway Borehole. In the latter, the top of the halite is between 26 and 47 m below the Arden Sandstone (depending on which green bed is correlated as the Arden Sandstone; see below), suggesting that the equivalent beds in the Worcester district lie within the palynologically barren upper part of the Worcester Borehole. Geophysical correlations (Figure 23) confirm this suggestion. The Droitwich Halite lies above beds equivalent to the Northwich Halite of the Cheshire Basin and is correlatable with the Wilkesley Halite of that area.

Arden Sandstone Formation

The Arden Sandstone Formation generally forms a prominent escarpment which is best developed in the north and around The Old Park [87 46]. It has a very disjointed outcrop in the south of the district, where it is broken by faulting and, to a lesser extent, folding. East of Worcester, the outcrop is continuous but traversed by a few small faults.

The formation is 11m thick in its type area in Warwickshire (Old et al., 1991). In this district, it thins northwards from 6.4 m in the Upton on Severn Borehole to 1 to 2 m at Brook End; to the south it is 4.5 m in the Twyning Borehole. East of Worcester it is 3 to 5 m thick in the Warndon area [895 571], thinning to 0.8 m about 1 km to the north [8961 5802].

North of the district the formation is absent at outcrop in the Droitwich area or represented by a different facies. It is, however, almost certainly present in the nearby Saleway Borehole, where there are four beds of dominantly green mudstone and siltstone between 210.92 and 246.89 m depth. Wills (1970, 1976) correlated the Arden Sandstone with the highest of these beds, (0.3 m of dark grey, shaly mudstone) but, alternatively, it may correlate with 7.85 m of green beds at 222.12 m.

The formation consists of interbedded mudstones, siltstones and fine-grained sandstones which are predominantly grey-green to pale grey in colour. The proportions of the constituent lithologies vary both laterally and vertically. The most westerly outcrop of the formation in the south of the district consists entirely of siltstone. In the Brook End–Hatfield area to the south-east of Worcester the formation is a red, medium- to fine-grained sandstone. This was originally assigned to a horizon lower in the Mercia Mudstone Group and named the Brook End Sandstone (e.g. Ambrose et al., 1985), but it is now correlated with the Arden Sandstone, on the basis of geophysical log data of the Kempsey Borehole and other boreholes (Figure 23).

The mapped boundaries of the formation are taken at the colour change from red to green. However, the boundaries do not correspond to the lithological change, as there is commonly a secondary green reduction zone in the overlying and underlying red mudstones, up to 0.3 m thick. Near Pirton [877 473] red mudstones occur within the formation.

The formation is mainly thinly bedded to laminated, with parallel lamination and lenticular bedding. The sandstones are generally fine grained and well sorted but locally coarse grained and poorly sorted. Mud-flake breccias occur locally at the base of some beds. Sedimentary structures include cross-bedding and cross-lamination, thixotropic soft sediment deformation and small-scale scours. Current directions measured at Sudeley Farm [8718 4042] are unimodal and from the south-west to south-south-west in one bed, but polymodal in others. Gypsum veins occur locally.

The formation has yielded a relatively diverse flora and fauna throughout the central and south Midlands, although only plant debris and *Euestheria* have been noted during the present survey. Recorded in neighbouring areas are possible marine bivalves, remains of fish (including sharks) labyrinthodont amphibians, and chirotheroid and rhynchosauroid tracks (Old et al., 1991). Burrows are common throughout the formation, particularly at sandstone–mudstone interfaces. Horizontal and vertical types are present, and in some beds, the bedding has been totally destroyed by bioturbation; *Planolites* and *Phycodes* types are typical. The formation has yielded palynological assemblages (Table 7) dominated by bisaccate pollen derived from gymnosperms, principally conifers, but including small numbers of representatives of the circumpolles group, and some spores of lycopsids, ferns and a bryophyte. The association of *Duplicisporites* spp., *Haberkornia gudati*, *Camerosporites secatus*, *Vallasporites ignacii*, *Ovalipollis pseudoalatus*, *Ellipsovelatisporites plicatus*, *Ricciisporites umbonatus*, *Patinasporites densus* and *Brodispora striata* is indicative of a Tuvalian (late Carnian; Late Triassic) age. The assemblages are comparable with those from the formation in the Tewkesbury district (Clarke, 1965; Warrington, 1970a) and from localities in the Redditch district (Old et al., 1991).

The faunal associations and the remains of the colonial chlorophyceaen alga *Botryococcus*, and of *Plaesiodictyon*, indicate dominantly subaqueous, fresh- to brackish-water conditions. Shark remains and a possible herkomorph acritarch (*Cymatiosphaera?*) indicate marine involvement in the creation of the brackish environments. The Arden Sandstone appears, therefore, to represent deposition in a range of subaqueous environments subjected to varying salinities, perhaps in an estuarine or similar setting (Warrington and Ivimey-Cook, 1992). The bioturbated argillaceous beds and sheet sandstones were probably deposited in interdistributary areas, the cross-bedded sandstones being the deposits of minor distributary channels.

Four good, but incomplete, sections of the formation were seen at the time of survey. The best was in a disused quarry at Sudeley Farm [8718 4042] near Ryall (Plate 12), which exposes a 3 m section comprising fine-grained, ripple-laminated and parallel-laminated sandstone above mudstones with thin sandstone layers. Two sections at Severn Stoke [8571 4363; 8577 4377] are mainly of mudstone, as is the moderately well-exposed section in the Blue Hill railway cutting [900 528], Spetchley.

Twyning Mudstone Formation

The Twyning Mudstone Formation is named from the BGS Twyning Borehole [8943 3664]. The formation forms the eastern part of the Mercia Mudstone Group outcrop. Faulting has produced an irregular outcrop, with downfaulted blocks, an inlier in the north-east at Crowle, and several small outliers in the south. The topography is similar to that of the Eldersfield Mudstone Formation, with subdued dip and scarp features. The general dip is to the east and south-east, with variations caused by the faulting and gentle folding.

The formation consists mainly of red, blocky mudstones. In the Twyning Borehole some units have traces of bedding and a few contain intraformational mudstone clasts. There are several mappable, more resistant, silty mudstone and siltstone beds which commonly contain numerous reduction spots. Laminated beds and green beds are comparatively rare.

Nodular gypsum/anhydrite is common throughout much of the formation, but is absent in the upper part. The evaporite-bearing beds include a geophysical marker in the Twyning and Netherton [9982 4137] boreholes (Figure 23) which correlates with the Spernall Gypsum of Warwickshire (Ambrose, in preparation). The uppermost beds of the formation are cut out to the north-east by an unconformity at the base of the Blue Anchor Formation (Old et al., 1987; Ambrose, in preparation), but there is no evidence for a break at this level in the Worcester Basin.

There are no accurate thickness measurements for the formation within the district, but the Twyning Borehole proved 169.01 m, the Netherton Borehole about 185 m and the Saleway Borehole at least 175 m. Field mapping suggests substantial thinning to the east and south-east of Worcester, but this is not supported by seismic reflection data (Figure 36). Greater throws on the mapped faults and possibly on other, undetected, strike faults may explain this disparity. The formation thins to the north-east and is about 50 m thick in the eastern part of the Redditch district (Old et al., 1991).

No fossils have been found in the Worcester and adjoining districts. The formation is assigned a Norian age, as it occurs between the proven Late Carnian Arden Sandstone Formation and the Rhaetian Blue Anchor Formation.

Blue Anchor Formation

The Blue Anchor Formation crops out mainly in the prominent north-south–trending escarpment in the east

Plate 12   Arden Sandstone Formation, Sudeley Farm [8718 4042] near Upton upon Severn. A 15071

of the district. The outcrop is broken by faulting around Crowle [92 56], Upper Wolverton [91 50] and Littleworth [88 49]. The escarpment dies out for a short distance around Littleworth where the formation is cut out by the Pirton Fault. A marked indentation of the escarpment between Upper Wolverton and Littleworth is caused by gentle folding.

The formation ranges in thickness from about 3 m between Abbots Wood [888 494] and Pirton [885 470] to 15 m at Thrift Wood [914 556], south of Crowle. Boreholes and sections in the district and neighbouring areas prove thicknesses between 8 and 11 m (Table 8). Thicknesses between 3 and 10 m were recorded in the Tewkesbury district (Worssam et al., 1989); in the Redditch area the formation thins locally to 0.1m, probably as a result of erosion and uplift along the Vale of Moreton Anticline (Old et al., 1991).

The formation consists of pale greenish grey siltstones and darker silty mudstones with muddy laminae and fine-grained sandstone layers and lenses. In the highest beds there are thin layers of hard, indurated, pale creamy buff porcellanous mudstone and siltstone. The colour change at the junction with the underlying red mudstones of the

Twyning Mudstone Formation varies from sharp and irregular to apparently transitional. At outcrop, the change from red to pale greenish clay soils is very marked.

Table 8   Thicknesses of the Blue Anchor Formation recorded in sections and boreholes in and adjacent to the Worcester district.

	Thickness m
Saleway Borehole [9285 6017][5,6,7]	9.14
Dunhampstead railway cutting [919 600][2,3]	10.4
Dunhampstead railway cutting [919 600][4]	10.7
Trench Wood railway cutting [918 598][2]	c.9.1
Climer's Hill [9183 5707][4]	c.9.4
Norton railway cutting [893 512][1]	10.7
Bourne Bank [890 420]	c.9.0
Baughton Hill motorway cutting [890 414]	8.1
Twyning Borehole [8943 3664][8]	9.75

1 Strickland (1842)	5 Wills (1970)
2 Harrison (1877)	6 Wills (1976)
3 Richardson (1904a)	7 Poole and Williams (1981)
4 George (1937)	8 Worssam et al. (1989)

Fish scales have been found in the Twyning (Worssam et al., 1989) and Knowle boreholes (Old et al., 1991). The latter also yielded sporadic miospores and dinoflagellate cysts. The Saleway Borehole (Table 9) yielded the conifer pollen *Geopollis zwolinskai*?, known from late Triassic, Norian? and Rhaetian sequences, and the colonial chlorophycean alga *Botryococcus*.

The formation is of lacustrine origin, the presence of desiccation cracks indicating periodic subaerial exposure.

PENARTH GROUP

Rocks of the Penarth Group cap the prominent north–south escarpment in the east of the district and form a nearly continous outcrop, broken in a few places by faulting. There are three inliers; one north-east [915 502] of Stoulton results from faulting and those south [926 540] of Broughton Hackett and north [940 582] of Huddington are due to small-scale valley bulging. In the north, easterly dips of 1° to 2° give rise to a broader outcrop than farther south where the dip is between 6° and 12°.

The group comprises 8 to 12 m of mainly shales and calcareous mudstones, with subordinate sandstones and thin limestones. In the lithostratigraphical classification of Warrington et al. (1980), the Penarth Group replaces the former terms 'Rhaetic Beds' and '*Avicula contorta* Shales'. The group is divided into an older Westbury Formation and a younger Lilstock Formation, the latter represented here only by the Cotham Member (but see also p.80). Richardson (1903a, 1903b, 1904a) described all the main sections in Worcestershire. He assigned bed numbers to the better exposures at Wainlode Cliff and Garden Cliff in Gloucestershire (1903a) and correlated these with the exposures in Worcestershire (Figure 25). Richardson (1904b, 1948, 1964) discussed the relationship of the 'Rhaetic Beds' with the overlying and underlying strata. He argued that they were laid down on an irregular surface of gently folded older Triassic beds, with onlap of successive sandstone and shale beds east and west of a major synclinal flexure southwards from Dunhampstead, north of the district. Earlier, Strickland (1842) and Brodie (1845) included these beds in the Lower Lias, and Wright (1860) separated, as the Zone of *Avicula contorta*, all those beds between the 'grey, green and red marls of the Keuper, and the lowest *Ostrea* Beds'. Harrison (1877) described in detail the sections at Crowle and Dunhampstead and George (1937) gave an account of the 'Rhaetic Beds' in the northern part of the district, as far south as Climers Hill [918 570], Crowle. The rocks of the Tewkesbury and Redditch districts have been recently described by Worssam et al. (1989) and Old et al. (1991) respectively.

The principal sections in the district are illustrated in Figure 25. The group maintains a similar thickness southwards into the Tewkesbury district (Worssam et al., 1989) and to the north and north-east, apart from in the Saleway Borehole (see below) and at Round Hill [SP 1427 6179], north of Stratford upon Avon, where only the Cotham Member is present (Old et al., 1991).

Palynomorph assemblages were recovered from the Saleway (Table 9) and Twyning (Table 10) boreholes and from the Cotham Member in a section at Broughton Hackett [9265 5431] (Table 11). The assemblages comprise associations of miospores (spores and pollen) of land plant origin, organic-walled microplankton (acritarchs, dinoflagellate cysts and tasmanitid (?) algae) and other remains indicative of marine environments. The miospore associations display a progressive change in composition and an increase in diversity upwards from the Westbury Formation into the Lilstock Formation. The palynomorph succession recorded in the Saleway Borehole (Table 9) is broadly comparable with those from the Twyning Borehole (Table 10) and others in the south Midlands. The occurrence of *Quadraeculina anellaeformis* and *Tsugaepollenites*? *pseudomassulae*, in association with *Ovalipollis pseudoalatus*, *Rhaetipollis germanicus* and *Ricciisporites tuberculatus*, indicates a Rhaetian (Late Triassic) age.

The Twyning Borehole provides a suite of geophysical logs through the group (Figure 26). There is an overall good match of the logs with the lithologies recorded from the core apart from the long spaced density (LSD) log. The self-potential (SP) log has a very subdued signature, only the 0.58 m sandstone near the base of the Westbury Formation causing any significant deflection. The remaining logs all have moderately serrated signatures, reflecting the variations within the sequence. The base of the group is best defined on the gamma-ray and SP logs, is reasonably well defined on the resistivity log, but poorly defined on the others. The Westbury Formation–Cotham Member junction is also poorly defined apart from on the sonic log, reflecting the transitional nature of this boundary.

Westbury Formation

The Westbury Formation ranges from 5 to 10 m thick, averaging 6 to 7 m. It comprises two units; the lower one, 2 to 3 m thick, consists of interbedded sandstones and dark grey to black, pyritic, fossiliferous shales. The sandstones are typically fine grained and micaceous, with cross-bedding, cross-lamination, wavy bedding, load casts and burrowed surfaces. They commonly contain bivalves and fish teeth, scales and bone fragments. The grains are dominantly angular, indicating a nearby source (George, 1937). The upper unit comprises 3 to 8 m of grey, dark grey and black, fissile and blocky mudstones and silty mudstones with subordinate siltstones and nodular, silty to sandy, shelly and shell detrital grainstones. The upper part is transitional with the overlying Cotham Member and consists of finely interlaminated grey and dark grey mudstone and siltstone. In the northern part of the district, the lower unit produces a pronounced gentle dip slope littered with sandstone debris and can be easily mapped. South from the Littleworth–Upper Wolverton area [89 51] the dip is steeper and the sandstones rarely form dip slopes.

The only complete sequence proved in the area was that formerly exposed in the M5 Motorway cutting at Baughton Hill [890 414–891 413], recorded by Dr H C

Table 9   Distribution and relative abundances of palynomorphs in the Blue Anchor Formation (Mercia Mudstone Group), Penarth Group and basal Lias Group, Saleway Borehole (for location see Figure 20).

MMG	PENARTH GROUP			LIAS GROUP	Lithostratigraphic units
Blue Anchor Formation	Westbury Formation	Lilstock Formation			

Depth in borehole (m): 25 — 20

Palynology sample horizons and preparation numbers (MPA): 19829, 19828, 19827, 19826, 19825, 19824, 19823, 19822, 19821, 19820

Key:
- ○ 0–1%
- ● >1–5%
- |‖‖‖| >5% (bar scale) 50%
- + Present (not counted)
- ? Questionable occurrence
- MMG Mercia Mudstone Group

Miospores:
Geopollis zwolinskai
Converrucosisporites luebbenensis
Leptolepidites argenteaeformis
Granuloperculatipollis rudis
Alisporites thomasii
Tsugaepollenites? pseudomassulae
Quadraeculina anellaeformis
Vesicaspora fuscus
Ricciisporites tuberculatus

Porcellispora longdonensis
Ovalipollis pseudoalatus
Rhaetipollis germanicus
Classopollis torosus

Protohaploxypinus cf. microcorpus
Concavisporites sp.
Alisporites sp.
Lunatisporites rhaeticus
Convolutispora microrugulata
Carnisporites anteriscus
Perinosporites thuringiacus
Deltoidospora neddeni
Convolutispora sp.
Abietineaepollenites dunrobinensis
Podocarpites sp.
Acanthotriletes ovalis
Kraeuselisporites reissingeri

Carnisporites lecythus
Limbosporites lundbladii
Retitriletes austroclavatidites
Taurocusporites sp. A
Microreticulatisporites fuscus
Acanthotriletes varius
Chasmatosporites magnolioides
Gliscopollis meyeriana
Indeterminate disacciatriletes

Organic-walled microplankton and other remains:
Micrhystridium sp.
M. lymense var. gliscum
Rhaetogonyaulax rhaetica

Dapcodinium priscum
Cymatiosphaera polypartita
Micrhystridium licroidium
M. lymense var. lymense
?Beaumontella caminuspina
Foraminifera (test linings)

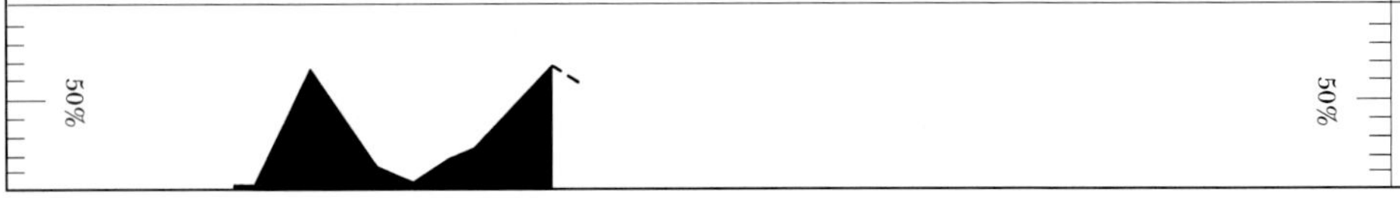

Relative abundance of miospores (white) to other palynomorphs (black)   50%

Relative abundances are expressed as percentages based upon counts of 200 specimens (except in MPA 19829 and 19821)

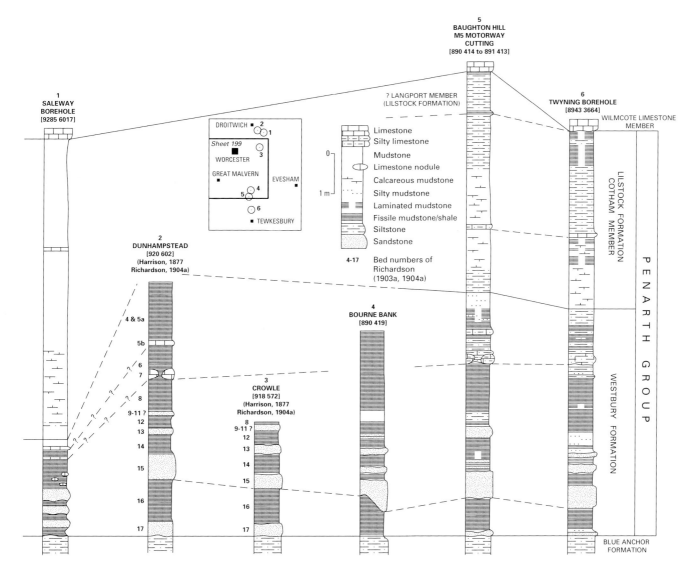

Figure 25 Sections of the Penarth Group in the Worcester and adjoining districts.

Ivimey-Cook in 1961. Here the Westbury Formation is 6.15 m thick, with 2.55 m of interbedded sandstone and shale at the base. The lowermost bed of the formation is either a sandstone ('Bed 17' of Richardson, 1903a) or shale ('Bed 16'), resting non-sequentially on the underlying Blue Anchor Formation. At Baughton Hill the base is scoured, the scours filled with up to 30 mm of fine conglomerate composed of quartz pebbles, fish fragments and coprolites. The most persistent sandstone is 'Bed 15' which Richardson (1903a, b; 1904a ,b) correlated with the 'Bone Bed' at Wainlode Cliff. There, it is a thin pyritic sandstone with fish debris and coprolite impressions, and passes laterally into a thicker, micaceous sandstone devoid of fish debris. However, the correlation of individual sandstones is tenuous in view of their lenticular nature and a variable number (from two to four) of beds present in sections (Figure 25). The small pebble conglomerate at the base of the lowermost bed at Baughton Hill, which apparently correlates with 'Bed 17' of Richardson, is similar to

the 'Bone Bed' at Wainlode and may correlate with 'Bed 15'. This facies may develop locally within any of the sandstones, and Richardson (1903a) recorded bone beds in the three lowermost sandstones at Garden Cliff, Westbury upon Severn, Gloucestershire. George (1937) found fish debris in sandstone fragments at several localities in the north of the district.

Higher in the sequence, there are two persistent nodular limestones ('Beds 5b and 7'), which form topographical features locally and can be mapped. Richardson (1903a) noted Bed 5b at Bourne Bank in a section adjacent to the one recorded during the recent survey (Figure 25). North of the district, the Saleway Borehole appears to prove a much attenuated sequence of the Westbury Formation (Figure 25). A reappraisal of the lithostratigraphy and palaeontology (Ambrose, 1984) has revised the position of the top of the Westbury Formation from that given by Poole and Williams (1981), suggesting that only about 2.3 m are present. Specimens

Table 10   Distribution and relative abundances of palynomorphs in the Blue Anchor Formation (Mercia Mudstone Group), Penarth Group and lower Lias Group, Twyning Borehole (for location see Figure 20)

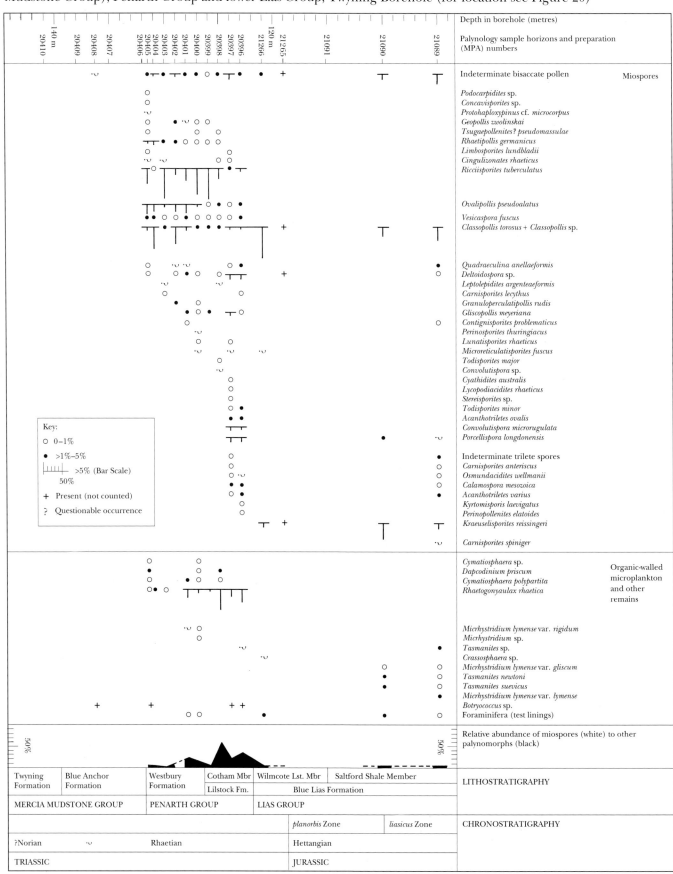

Relative abundances are expressed as percentages based upon counts of 200 specimens (except in MPA 20408 and 21265)

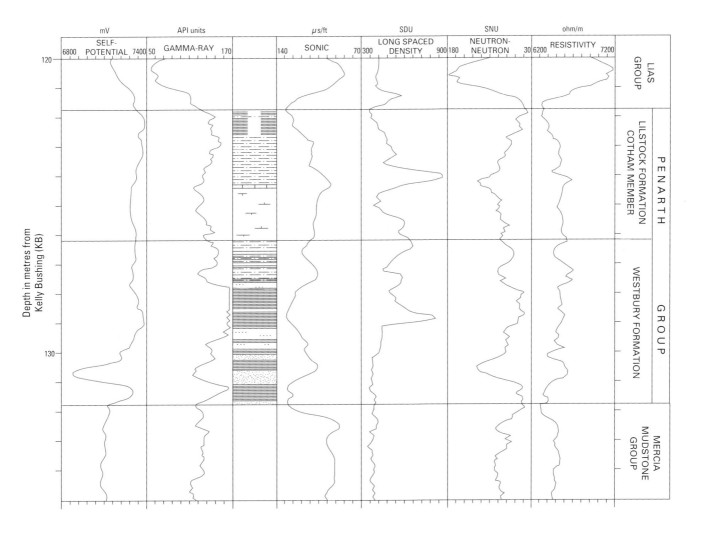

Figure 26   Composite log of the Penarth Group of the Twyning Borehole. See Figure 25 for key to lithologies.

collected suggest that the junction with the overlying Cotham Member is apparently transitional, with interbedding of the two characteristic lithologies. The revised thickness of the Cotham Member is greater than that proved over the rest of the district. Outcrop data 500 m to the south-east suggest a normal sequence and the depths given in the borehole are therefore considered unreliable. A possible attenuated sequence was observed during the recent survey in a temporary pipe trench at Shell [9514 5977–9515 5982] in the extreme south-west of the Redditch district. Here, typical Cotham Member mudstones rest on black shales of the Westbury Formation with no intervening transition beds, suggesting a non-sequence between the two formations.

The Westbury Formation contains a rich palynomorph association (Tables 9, 10) and a fauna of marine bivalves, gastropods, fish and echinoid debris. Figure 27 shows the macrofauna collected from the Baughton Hill section. The most common fish remains are *Acrodus*,

Gyrolepis and *Saurichthys*. Fauna from other localities is given in George (1937), Harrison (1877) and Richardson (1903a, b; 1904a).

Lilstock Formation

Cotham Member

The Cotham Member comprises 3 to 5 m of pale and medium greenish grey, commonly blocky, calcareous mudstones with a few thin siltstones and nodular wackestones. Brown calcareous mudstone beds occur locally. The upper part is finely interlaminated with siltstone in some areas. The member everywhere forms a narrow outcrop, usually at the base of a small scarp slope formed by the overlying Wilmcote Limestone. The base is transitional from the underlying Westbury Formation and is drawn at the incoming of calcareous mudstones. The top is sharp in unweathered sections and is marked by a non-sequence. Locally,

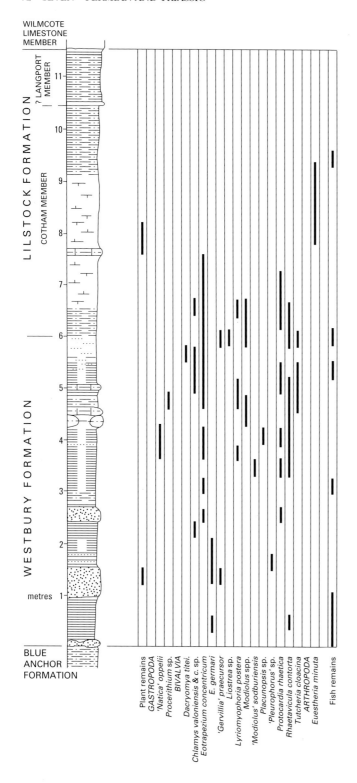

Figure 27 Fauna from the Penarth Group, Baughton Hill cutting. See Figure 25 for key to lithologies.

grey, calcareous mudstones, typical of the Lias, occur at the base of the overlying Wilmcote Limestone (see Chapter 8). In the weathered state they cannot be distinguished easily from Cotham Member mudstones and the boundary has been taken at the incoming of limestone.

The only complete sequence of the Cotham Member proved in the district is the M5 Motorway cutting at Baughton Hill (Figure 25) where it is 4.35 m thick. The Twyning Borehole proved a similar sequence, 4.46 m thick. In the Saleway Borehole, the Cotham Member appears to be at least 4.7 m thick and possibly up to 7.5 m thick. The uppermost 2.90 m of beds (between the depths 19.56 and 16.66 m), assigned to the Lower Lias by Poole and Williams (1981), may also belong to the Penarth Group but no samples were collected.

The Cotham Member is largely devoid of macrofossils apart from the locally abundant crustacean *Euestheria minuta*, a few fish fragments, ostracods and plant remains. A few bivalves were collected from the lowermost beds at Baughton Hill (Figure 27). The member has a rich and varied palynomorph assemblage (Tables 9, 10, 11).

The Penarth Group represents the widespread establishment of a shallow-marine, shelf sea containing a diverse fauna. The black shales of the Westbury Formation indicate anaerobic conditions on or below the sea floor and the sandstones probably represent migrating sand bars. Later, the sea became more aerated, with some influxes of arenaceous sediment and periodic carbonate precipitation. During deposition of the Cotham Member, carbonate precipitation was widespread, and took place in shallow, fresh and brackish water, as indicated by *Euestheria* and darwinulid ostracods, and marine conditions, as indicated by the presence of marine palynomorphs.

Table 11 Distribution and relative abundances of palynomorphs in the Cotham Member (Lilstock Formation, Penarth Group) and Wilmcote Limestone member (Blue Lias Formation, Lias Group), Broughton Hackett.

BED NUMBERS

2   3   4   5   6   7   8   9

SAMPLED SECTION

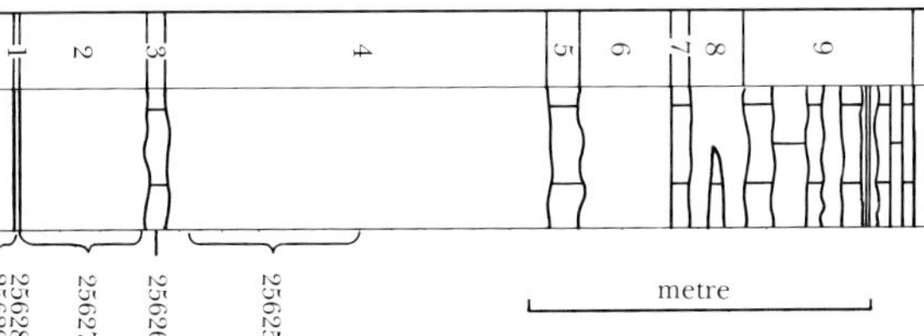

PALYNOLOGY SAMPLE HORIZONS AND PREPARATION (MPA) NUMBERS

MPA numbers: 25630 · 25628 / 25629 · 25627 · 25626 · 25625

metre

Key

○	0–1%
●	>1–5%
‖‖‖	>5% (Bar Scale)
	50%
?	Questionable occurrence
cf.	Comparable form present

Miospores

Concavisporites sp.
Triancoraesporites reticulatus
Vesicaspora fuscus
Alisporites spp.
Rhaetipollis germanicus
Convolutispora microrugulata
Kyrtomisporis laevigatus
Limbosporites lundbladii
Chasmatosporites magnolioides
Lunatisporites rhaeticus
Ovalipollis pseudoalatus
Cingulizonates rhaeticus
Perinopollenites elatoides
Carnisporites spiniger
Gliscopollis meyeriana
Acanthotriletes ovalis
Carnisporites anteriscus
Indeterminate trilete spores
Deltoidospora sp.
Calamospora mesozoica
Acanthotriletes varius
Lycopodiacidites rugulatus
Todisporites minor
Indeterminate bisaccate pollen
Ricciisporites tuberculatus
Classopollis torosus + Classopollis sp.

Cornutisporites seebergensis
Zebrasporites laevigatus
Polycingulatisporites bicollateralis
Stereisporites sp.
Triancoraesporites ancorae
Zebrasporites interscriptus
Leptolepidites argenteaeformis
Densosporites sp.
Porcellispora longdonensis
Contignisporites problematicus
Converrucosisporites cameroni
Quadraeculina anellaeformis
Microreticulatisporites fuscus
Stereisporites radiatus
Punctatisporites sp.
Carnisporites leviornatus
Kraeuselisporites reissingeri

Nevesisporites bigranulatus
Converrucosisporites luebbenensis
Camarozonosporites cf. rudis
Acanthotriletes sp.
Osmundacidites wellmanii

Organic-walled microplankton and other remains

Dapcodinium priscum
Tasmanites sp.
Rhaetogonyaulax rhaetica
Beaumontella langii
Micrhystridium lymense var. gliscum
?Beaumontella caminuspina
Micrhystridium sp.
Cymatiosphaera sp.
Foraminifera (test linings)

Relative abundance of miospores (white) to other palynomorphs (black)

50%   50%

LITHOSTRATIGRAPHY

Cotham Member	Wilmcote Limestone Member
Lilstock Formation	Blue Lias Formation
Penarth Group	Lias Group

Relative abundances expressed as percentages based upon counts of 200 specimens.

EIGHT

Jurassic

Jurassic rocks crop out east of the prominent escarpment formed by the underlying rocks of the Penarth Group. They are represented by the Lias Group and lowermost part of the Inferior Oolite Group, the latter being present only in an outlier capping Bredon Hill in the extreme south-east of the district. The Lias Group rocks strike north–south, with an easterly dip. Minor folding and faulting causes an indentation of the escarpment at Upper Wolverton [91 50]. The lowest beds form a small subsidiary escarpment immediately east of the main Penarth Group escarpment and extensive dip slopes in the north; south of Upper Wolverton steeper dips result in much narrower dip slopes. To the east is a narrow clay vale underlain by the Saltford Shale and a much less prominent escarpment, formed by the Rugby Limestone. In the south-east dips decrease and the Lias Group forms predominantly low lying, gently undulating terrain with flat-topped hills, some of which are capped by river terrace deposits. The River Avon flows across the main clay vale in the south-east of the district, beyond which the ground rises steeply to Bredon Hill. The higher beds of the Lower Lias, the overlying Dyrham Siltstone Formation, Marlstone Rock Formation, Upper Lias and lowermost Inferior Oolite Group rocks are all largely obscured by landslip deposits on the flanks of Bredon Hill.

The Lias Group consists mainly of blocky and fissile calcareous mudstones, with subordinate limestones. In the lower part there are two mappable units of interbedded limestone and mudstone in the Blue Lias Formation, the Wilmcote Limestone and Rugby Limestone members; higher in the sequence are siltstones (the Dyrham Siltstone Formation) and sandstones and sandy limestones (the Marlstone Rock Formation). The group is estimated to be around 500 m thick, perhaps thickening to around 700 m at Bredon Hill (Whittaker, 1972a, b). To the south, there are 450 m in the Tewkesbury district (Worssam et al., 1989) and 496.8 m in the Stowell Park Borehole [SP 084 118], north-east of Cirencester (Green and Melville, 1956). Eastwards the group thins to around 270 m in the Warwick district. At least 22 m of oolitic limestones of the Inferior Oolite Group overlie the Lias Group.

Much of the previous work in the district focussed on the lowest beds, which were quarried locally in the 19th century. A number of quarry sections were published and several stratigraphical classifications proposed (Table 12). Worssam et al. (1989) retained the existing Lias Group nomenclature in the Tewkesbury district, Old et al. (1991) introduced a revised lithostratigraphy for the Blue Lias Formation. The highest beds, present on Bredon Hill, have been described by Buckman (1903), Richardson (1902, 1905, 1924, 1929), Whittaker (1972b), Williams and Whittaker (1974), Worssam et al.

(1989), and Simms (1990). More general references to the Jurassic rocks of the district include Arkell (1933) and Woodward (1893, 1894).

DEPOSITIONAL HISTORY

The sediments were deposited in a shallow-marine environment in a warm, humid climate. The nearest land was probably the London Platform to the south-east, part of the Anglo-Brabant Landmass (Bradshaw et al., 1992). The presence of land to the west is unproved, but, farther south, the margin of the South Wales Coalfield was being eroded in earliest Jurassic times. Eastern Worcestershire is one of the more distant areas from the London Platform of the present Lias outcrop of Central England. During marine transgression, successively younger beds onlapped onto the platform, so that the district became progressively more distal (Donovan et al., 1979). The proximity to a shoreline during deposition of the earliest Lias Group strata is shown by the remains of insects, terrestrial reptiles and plants. A mixed clastic–carbonate sequence comprises mudstones, muddy limestones and shell-fragmental limestones, the last produced by winnowing of fine sediment by current activity and storms. Bituminous paper shale was deposited in shallow, anaerobic conditions with minimal water circulation. Slight deepening of the water increased circulation and produced oxygenated waters (Hallam, 1967; Old et al., 1987).

The top of the Blue Lias Formation, here of earliest Sinemurian, *bucklandi* Zone age, marked the start of a long period of a more stable deeper water environment. At the close of the Lower Pliensbachian (*davoei* to *margaritatus* Zone) increased sedimentation resulted in shallower-water deposition of the silts of the Dyrham Siltstone Formation, and of the sands and sandy carbonates of the Marlstone Rock Formation. Ironstone, characteristic of the Marlstone Rock Formation elsewhere, is absent in this area, possibly indicating a more distal environment.

The Toarcian is marked by renewed transgression and a return to deposition of deeper-water mudstones of aerobic and anaerobic facies. A minor regressive event may have occurred in *H. falciferum* Subzone times, resulting in predominantly silt deposition. Bioturbation, as seen in the Bredon Hill No.1 (Lalu Barn) Borehole [SP 9577 3996] (Whittaker, 1972b), may indicate shallow-water deposition. Later in the Toarcian, a second regressive event produced the Cotteswold Sands. The base of these beds is diachronous, being of *variabilis* to *levesquei* Zone age at Bredon Hill, compared to a *bifrons* Zone age in the Stroud area. The youngest Toarcian beds, of *variabilis* to *levesquei* Zone age, are missing from most of the surrounding areas, in the east and south Midlands and the London Platform. The source of the sands is therefore not clear.

Clastic sedimentation continued into earliest Aalenian times, followed by the establishment of a widespread carbonate shelf, on which high-energy oolites and bioclastic limestones were deposited, with intermittent influx of terrigenous sands. Mudge (1978) interpreted the Bredon Hill succession as being initially intertidal to subtidal, passing successively to open shelf sand and oolitic shoal facies.

The Jurassic outcrop area lies within the central graben of the Worcester Basin, which is defined by major growth faults (Chapter 10), and there is a thicker sequence here than to the east. Contemporaneous movement occurred along the Lickey End (Inkberrow) Fault which bounds the Worcester Basin to the east of the district (Old et al., 1991). This may have influenced sedimentation patterns during the early part of Lias Group deposition. Movement in late Rhaetian–early Hettangian times may have resulted in localised areas of deeper water west of the fault, to which lime deposition was restricted. Later, minor inversion caused by reactivation of the same fault resulted in erosion prior to the deposition of the Marlstone Rock Formation and during the Aalenian (Simms, 1990).

GEOPHYSICAL LOG CHARACTERISTICS

Wireline logs are available from the Twyning Borehole, 3 km to the south of the district (Figure 28), and from the Bredon Hill No. 1 Borehole close to the south-east (Figure 29). In the former, they span the lowermost

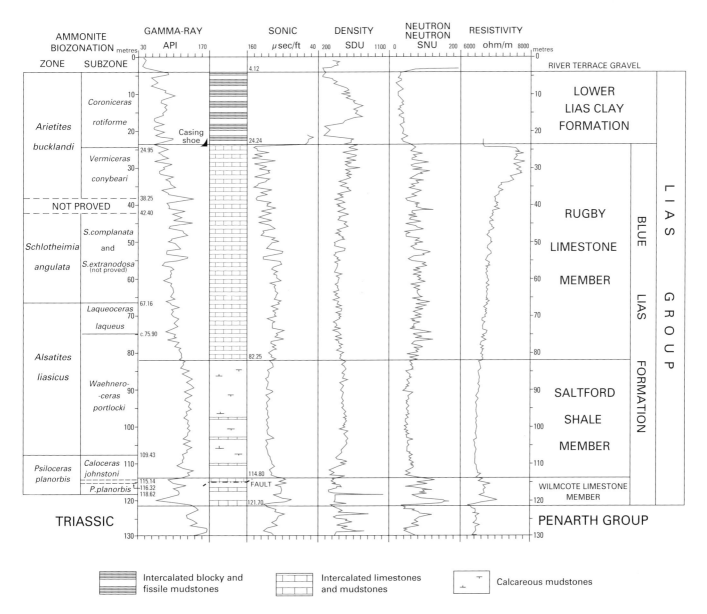

Figure 28 Composite log of the Lias Group in Twyning Borehole.

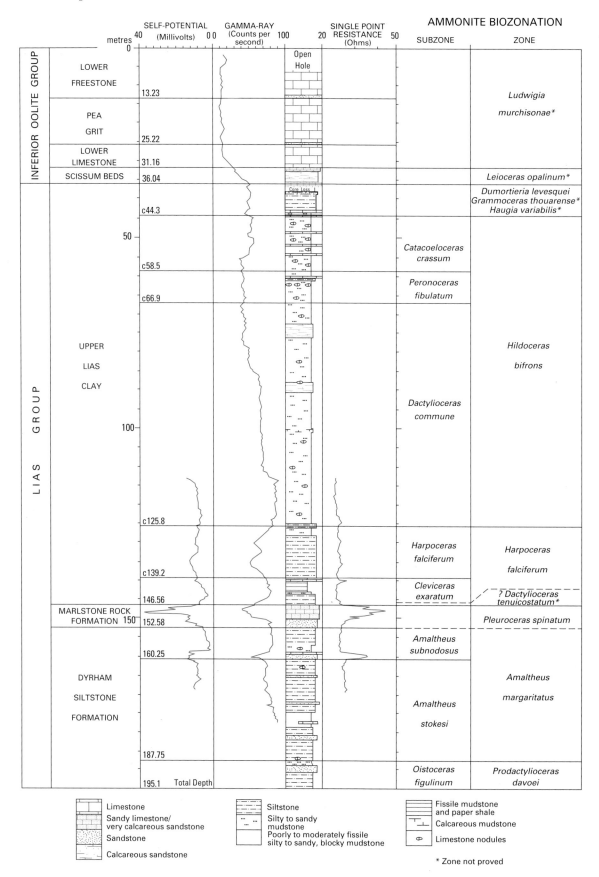

Figure 29   Composite log of the Bredon Hill No.1 (Lalu Barn) Borehole (revised from Whittaker, 1972b).

117.58 m of the Lias Group, including all of the Blue Lias Formation; the latter proved the highest beds from the Inferior Oolite Group down to a level within the Dyrham Siltstone Formation (*margaritatus* Zone). The lowermost 17.6 m of Bredon Hill No. 1 Borehole were not geophysically logged. Absolute values for the Self Potential (SP) and resistance logs of the Bredon Hill No. 1 Borehole are not available.

Whittaker et al. (1985) described and illustrated the characteristic Lias Group log signatures in the Wessex Basin, East Midlands Shelf and Yorkshire Basin. They identified nine subdivisions, LL1-5 (Lower Lias), ML1-2 ('Middle Lias') and UL1-2 (Upper Lias). Only LL1 is represented in the Twyning Borehole and ML2, UL1-2 in the Bredon Hill No.1. Borehole.

The lowest two members of the Blue Lias Formation produce characteristic geophysical signatures (Figure 28). The Wilmcote Limestone gives rise to a prominent set of peaks and troughs in all logs, although it is not well defined on the density log. The Saltford Shale Member produces subdued signatures, with higher gamma-ray, neutron-neutron and density values, and lower sonic velocity and resistivity values. The base of the Rugby Limestone Member is best defined by the neutron and resistivity logs. All of the logs show a distinct spiky signature in this member, reflecting the interbedding of limestones and mudstones. The top of the Blue Lias cannot be defined from the gamma-ray log, the spikiness of the signature continuing to the top of the borehole in the overlying 20.12 m of Lower Lias Clay. Other logs are affected by the presence of casing down to a level which corresponds approximately with the top of the Blue Lias. The density log, however, shows a change in the character of its signature with greater variations in density values. A comparison of the Blue Lias gamma-ray log of Twyning with those of other boreholes from the Wessex Basin (see Whittaker et al., 1985) shows that the Blue Lias Formation as defined by their unit LL1 continues above the lithostratigraphic top defined from the core in Twyning. The continued spikiness of the Twyning log suggests alternations of lime-rich and lime-poor mudstones, not identified in the core; the former may be laterally equivalent to limestones elsewhere in the basin.

The gamma-ray log of the Bredon Hill No.1 Borehole probably commences in unit ML2 of Whittaker et al. (1985), although the lowermost part with the high gamma-ray values may be the top of ML1. Overall there is a very good match of lithology with the wireline signatures. The Marlstone Rock Formation is marked by a pronounced gamma-ray low, high SP and resistance values. It is best defined by the SP log, whereas the upper interbedded limestones and sandstones are most clearly defined on the single point resistance log (Figure 29). In the Upper Lias, the gamma-ray signatures are generally subdued. The siltstones of the *H. falciferum* Subzone give rise to a marked gamma-ray low within the UL1 division. There is a pronounced interval of low gamma-ray values within the Upper Lias, for which there is no obvious explanation from examination of the borehole core description. The inflection at the start of this decrease in

radioactivity is taken as the boundary between UL1 and UL2. The boundary between the Lias and Inferior Oolite groups is clearly marked by a rapid decrease in gamma-ray values, although the base of the Cotteswold Sands at the top of the former and the subdivisions of the latter are poorly defined.

STRATIGRAPHY

The lithostratigraphy and biostratigraphy of the Jurassic strata of the district are summarised in Table 12. The Liassic ammonite zones and subzones follow Dean et al. (1961) and Cope et al. (1980a). The subzones in the *bucklandi* and *semicostatum* zones and the zone boundaries follow Ivimey-Cook and Donovan *in* Whittaker and Green (1983). The Aalenian ammonite zones follow Cope et al. (1980b). The base of the Jurassic System is marked biostratigraphically by the first appearance of the ammonites of the genus *Psiloceras* (Cope et al., 1980a). Since this fossil appears a few metres above the base of the Lias Group in this district, the beds underlying this horizon are assigned a Triassic age. They are, however, described here with the remainder of the Lias Group.

LIAS GROUP

Virtually all of the Jurassic rocks within the district belong to the Lias Group. The lithostratigraphical nomenclature for the Blue Lias Formation is that of Old et al. (1991) (Table 13). There is no suitable type section in the district to define a new formation for the overlying strata and the term Lower Lias Clay is retained, following Worssam et al. (1989). The term Middle Lias is abandoned, because of its biostratigraphically defined base, in favour of the Dyrham Siltstone Formation (Dyrham Silts of Cave, 1977). This is overlain by the Marlstone Rock Formation and Upper Lias Clay Formation (Table 12).

Blue Lias Formation

The Blue Lias Formation accounts for most of the Lias Group outcrop within the district. The three component members are easily recognised in the field, although the top of the formation is less well defined.

Wilmcote Limestone Member

The Wilmcote Limestone Member consists of interbedded limestones, mudstones and paper shales and includes the Stock Green Limestone at the base (Old et al., 1991; Figure 30). It produces a pronounced dip slope which is commonly littered with limestone debris. The thickness ranges from 1 to 2 m in the north to around 8 m in the south. The Twyning Borehole proved at least 6.9 m, with an unknown, but probably small thickness, faulted out. The member is consistently around 8 m thick across the Tewkesbury district to the south (Worssam et al. 1989) and is 6 to 8 m thick to the

Table 12 Jurassic stratigraphy.

	Stage		Ammonite Biozone	Lithostratigraphy of the Worcester district			Previous names
Jurassic	Aalenian (part)		*Ludwigia murchisonae*		Lower Freestone Pea Grit Lower Limestone	Inferior Oolite Group 25 m	Inferior Oolite Series
			Leioceras opalinum		Scissum Beds		
	Toarcian	Upper	*Dumortieria levesquei* *Grammoceras thouarsense* *Haugia variabilis*	'Cotteswold Sands' 7.7 m	Upper Lias Clay Formation 75–100 m	Lias Group 500 m	Upper Lias
		Lower	*Hildoceras bifrons* *Harpoceras falciferum* *Dactylioceras tenuicostatum*				Middle Lias
	Pliensbachian	Upper	*Pleuroceras spinatum*		Marlstone Rock Formation 6 m		Lower Lias
			Amaltheus margaritatus		Dyrham Siltstone Formation 60 m		
			Prodactylioceras daveoi				
		Lower	*Tragophylloceras ibex*		Lower Lias Clay Formation 250 m		
			Uptonia jamesoni				
	Sinemurian	Upper	*Echioceras raricostatum*				
			Oxynoticeras oxynotum				
			Asteroceras obtusum				
		Lower	*Caenisites turneri*				
			Arnioceras semicostatum				
			Arietites bucklandi	Rugby Limestone Member 55 m	Blue Lias Formation 85 m		
	Hettangian		*Schlotheimia angulata*				
			Alsatites liasicus	Saltford Shale Member 25–30 m			
			Psiloceras planorbis				
Triassic	Rhaetian (part)			Wilmcote Limestone Member 1.5–8 m			
						Penarth Group	Rhaetic

north-east in the Redditch district (Old et al., 1991). The localised thinning of the member in the north-east of the district is attributed to lateral facies change resulting from the earlier onset of slightly deeper-water sedimentation. This is further suggested by the presence locally of 1 to 2 cm of pebbly, shelly mudstone at the base of the member, seen in a temporary section [9265 5431] at Broughton Hackett (Figure 30). This mudstone is absent at Twyning and in the Evesham–Wilmcote area to the east.

The proportion of limestone is about 45 to 50 per cent in the Stock Green Limestone, and about 15 to 20 per cent in the overlying beds. Most of the limestones appear to be primary carbonate deposits. However, undulating bed boundaries and nodular layers suggest there may have been some diagenetic migration of lime.

Correlation of individual limestones within the Wilmcote Limestone is usually possible only over short distances (Figure 30), although a detailed correlation can be made between the sections at Stock Green in the Red-

Table 13 Lithostratigraphy of the Blue Lias Formation.

		This account Old et al. (1991)	Williams and Whittaker (1974)	George (1937)	Wright (1878)	Wright (1860)		Richardson (1904a, 1906)			Murchison (1845)	Others				Old and others (1987)	Edmonds et al. (1965)
Jurassic	Blue Lias Formation	Rugby Limestone Member	Blue Lias	Shales and limestones	Bucklandi Beds	Lima Beds	Bucklandi Beds	angulata–Beds	Limestones and shales	Plagiostoma Beds		Lima Beds (Brodie, 1865, 1868, 1874, 1875)				Blue Lias	Blue Lias
		Saltford Shale Member	Clays below the Blue Lias	Psiloceras shales	planorbis Beds	Ammonites planorbis Beds		Planorbis Beds								Lower Lias	Clays between the White Lias and Blue Lias
Triassic		Wilmcote Limestone Member	'basal Blue Lias Limestones'	Pre-planorbis or Ostrea Beds	Ostrea or Saurian Beds	Ostrea Beds Ostrea liassica Beds Ostrea Series Saurian Beds		Ostrea Beds Pre-planorbis beds Saurian beds	Lower Lias	Saurian Beds	Strensham Series (Judd, 1875)	Saurian Beds Brodie (1865)	Insect and Saurian Beds Brodie (1868)				

Broken lines indicate imprecisely defined boundaries

ditch district and Broughton Hackett, 7.5 km apart. Beds 3 and 5 are particularly distinctive, the former being a hard, dark grey, locally finely laminated, lime mudstone and the latter a pale grey, soft, lime mudstone ('cement-stone'). These lithologies have not been recognised with certainty to the south, but the basal bed at Strensham may correlate with Bed 3 at Broughton Hackett. No distinctive lithologies were identified in the core from the Twyning Borehole. The group of limestones assigned to the Stock Green Limestone is easily traceable across the district. Individual limestones were named by quarrymen in many of the former pits, particularly in the extreme south of the area (Brodie, 1845; Murchison, 1845; Wright, 1860). However, the exact location of many of the pits is not now known. Brodie (1845) referred to the lowest one or two limestones of the Lias Group in Gloucestershire and Worcestershire as the 'Insect Beds or Limestones', and these were correlated by Worssam et al. (1989) with the lowest two limestones in the Twyning Borehole (Figure 30).

The limestones are generally fine grained, argillaceous and flaggy; some are laminated. The lowest beds are shelly or shell-detrital packstones/grainstones, interbedded with lime mudstones. The shelly layers were probably storm generated. Higher beds are predominantly lime mudstones and contain much less shell debris. Thin sections of the Stock Green Limestone of the Redditch district have been described by Old et al. (1991), and similar lithologies are present in the Worcester district.

The biostratigraphical divisions of the Wilmcote Limestone are shown in Table 13 and Figure 30. The zonal boundaries are based on the correlation established by Old et al. (1991) and little additional data were obtained in the present district. In the north, the Wilmcote Limestone is entirely Triassic in age, and represented only by the Stock Green Limestone. South of Broughton Hackett, limestones appear at successively higher levels and the member is diachronous, extending up into the

planorbis Zone and Subzone. To the south of the district, the Twyning Borehole proved the top of the member to be in the *johnstoni* Subzone of the *planorbis* Zone.

The Wilmcote Limestone has yielded a rich and varied fauna with corals, gastropods, abundant bivalves, (particularly *Liostrea hisingeri*), ammonites, ostracods, echinoid and crinoid debris, fish and reptile remains, and insects. Faunal lists from various localities within the district were given by Brodie (1845), George (1937), Murchison (1845), Paris (1908), Richardson (1906), Woodward (1893) and Wright (1860). A number of plant remains from Strensham were named and illustrated by Buckman (1850).

Palynomorph assemblages from the Wilmcote Limestone have been recovered from a section at Broughton Hackett [9265 5431] (Table 11) and from the Saleway Borehole to the north and Twyning Borehole (Table 10). The assemblages compare closely with others from the basal part of the Lias Group across the Midlands. In the Redditch district, miospores recovered from the beds beneath the Stock Green Limestone bear some resemblance to those commonly found in the Langport Member, which overlies the Cotham Member of the Lilstock Formation (Old et al., 1991). The section at Broughton Hackett exposed the same part of the sequence (beds 1 and 2, Figure 30). These beds, together with the lower part of bed 4 (Table 11), yielded palynomorph assemblages comprising miospores and organic-walled microplankton. The presence of *Ricciisporites tuberculatus* and large numbers of *Kraeuselisporites reissingeri* in these beds indicate that they are stratigraphically equivalent to the Langport Member. The latter spore typically shows an abrupt increase in abundance immediately above the Cotham Member, for example in boreholes at Stockton Locks (Old et al., 1987), near Warwick, and Withycombe Farm (Warrington, 1978), near Banbury. The palynological record from Broughton Hackett supports the view (Donovan et al., 1979; Warrington and Ivimey-Cook, 1992) that the largely argillaceous beds

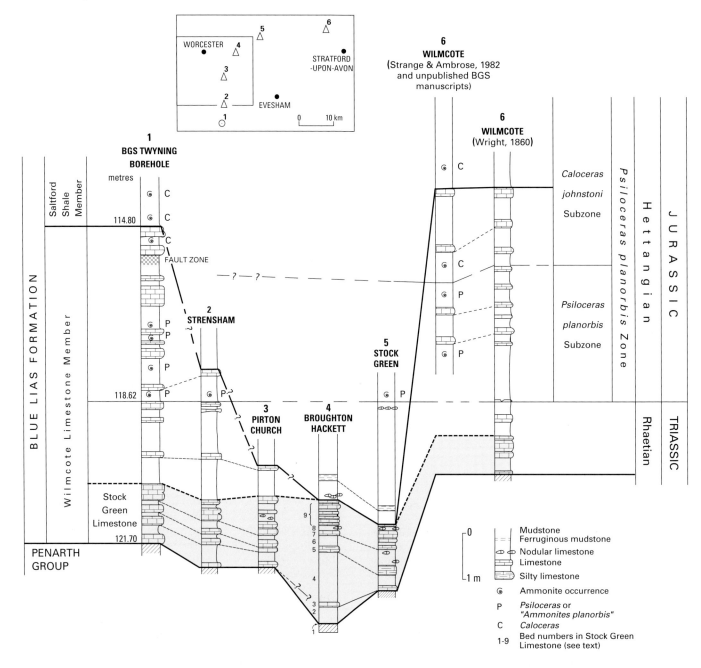

Figure 30   Comparative sections of the Wilmcote Limestone in the Worcester district and adjoining areas.

SALTFORD SHALE MEMBER

The Saltford Shale consists of dark grey mudstone with subordinate paper-shales and a few thin beds of limestone, cropping out in a narrow vale between escarpments formed by the Wilmcote Limestone and Rugby Limestone. Its thickness increases southwards from about 25 to 30 m, with 32.55 m proved in the Twyning

which succeed the Cotham Member in parts of central England are at least partly equivalent to more calcareous beds in the Langport Member (White Lias) farther to the east and south-east, towards the London Platform.

Borehole. There is also a southward facies change, with limestones included in the Wilmcote Limestone occurring at successively higher levels and increasing the thickness of this member at the expense of the Saltford Shale. The mudstones are fissile to blocky, fossiliferous and locally pyritic. At Stock Green, to the north-east of the district, fissile mudstone and paper shale are common in the lowest 6 m (Ambrose, 1984), whereas at Broughton Hackett the basal 1.4 m of the member, as seen in a temporary section [9265 5431], are of blocky mudstone. The few limestones present are either flaggy, shelly, shell detrital grainstones similar to those of the

Wilmcote Limestone, such as the beds cropping out near Stoulton [9107 5037], or rubbly weathering lime mud-stones ('cementstones') similar to those of the Rugby Limestone, such as the beds cropping out south-east of Crowle [9357 5550], north-east of Peopleton [93 50] and north-east [913 444] of Defford.

The member ranges in age from late Rhaetian to Het-tangian (late *liasicus* Zone). Beds of Rhaetian age are found only in the north, where up to 2.6 m of mudstone underlie the lowest occurrence of *Psiloceras planorbis,* as at Stock Green in the Redditch district (Figure 30). Southwards, the base of the member is diachronous and in the Twyning Borehole is within the *johnstoni* Subzone (Table 12). The age of the top of the member has not been proved in the north. In the higher parts of the se-quence, the common occurrence of the bivalve *Cardinia ovalis* suggests a *liasicus* Zone age. At Twyning the top is within the *liasicus* Zone, probably near the *portlocki* to *laqueus* subzonal boundary. Variations in the age of the top of the member may occur across the district, similar to those in the Harbury–Rugby area in the Warwick dis-trict, where it locally extends up into the *angulata* Zone (Old et al., 1987).

The fauna of the Saltford Shale is poorly known owing to the lack of boreholes and outcrop sections. Bivalves, particularly *Cardinia,* are common. Other fossils include the ammonites *Psiloceras planorbis, Caloceras* sp. and *Waehneroceras* sp., together with *Pentacrinites* and the trace fossil *Chondrites.* The following palynomorphs were recovered from a sample, probably of Saltford Shale, col-lected from a temporary exposure [9476 6077] to the north of the district: *Kraeuselisporites reissingeri, Quadrae-culina anellaeformis,* indeterminate non-taeniate bisaccate pollen, *Classopollis* sp., *C. torosus* and tasmanitid algae. Dr G Warrington reports that the assemblage is comparable with those known elsewhere from the lower beds of the Lias Group. The miospore component is indicative of a pre-Sinemurian age but is insufficient to determine whether the rocks are of Late Triassic (Rhaetian) or Early Jurassic (Hettangian) age.

RUGBY LIMESTONE MEMBER

The Rugby Limestone Member consists of alternating grey, lime mudstones ('cementstones') and dark grey mudstones, and is synonymous with the 'Blue Lias' of the Rugby–Harbury (Old et al., 1987) and Tewkesbury (Worssam et al., 1989) districts. The member forms a small scarp, its base easily recognised in the field by the appearance of pale grey rubbly lime mudstone, which is distinct from the flaggy limestones of the Wilmcote Limestone. Elsewhere at outcrop, lime mudstone debris is sporadic in fields and generally only seen in ditches. The top of the member is therefore difficult to locate precisely, although a distinctive limestone crowded with ossicles and columnals of the crinoid *Pentacrinites* is local-ly present near the top.

Limestone forms about 16 per cent of the member in the Twyning Borehole and this is probably typical throughout the district. Eastwards, the proportion of limestone increases to around 35 per cent in the Har-bury area (Old et al., 1987), reflecting reduced sedimen-

tation rates closer to the London Platform. Hallam (1964) considered the limestones to have formed both as primary carbonates and diagenetically.

The member is estimated to be about 55 m thick throughout the district, and 58.01 m were proved in the Twyning Borehole, which provides the best section of the member. The rhythmic sedimentation which charac-terises strata of this age in other areas (Hallam, 1960; Old et al., 1987) is poorly developed at Twyning. Bitumi-nous paper shales are rare, and restricted to a 7 cm bed at 27.95 m depth. Near-permanently oxygenated waters predominated in the district during deposition of the member, with bituminous muds accumulating in shallow waters closer to the shore of the London Platform, as in Warwickshire.

Few ammonites have been collected from the member within the district. In the Twyning Borehole, the mem-ber ranges in age from *liasicus* Zone (near the *portlocki–laqueus* Subzone boundary) through the *conybeari* Sub-zone of the *bucklandi* Zone and just into the earliest part of the *rotiforme* Subzone of the *rotiforme* Zone. Within the district, *Coroniceras* sp. and other arietitids, either of *buck-landi* or *semicostatum* Zone age, have been found at sev-eral localities. The zonally diagnostic *Schlotheimia angu-lata* was found east of Besford [9233 4543], indicating the *angulata* Zone. To the east of the district, *Caloceras* sp., which ranges from the *johnstoni* Subzone of the *planorbis* Zone into the *liasicus* Zone, was found at Grafton Flyford [962 562] and *Schlotheimia* sp. and *S. angulata,* indicating the *angulata* Zone, were found just south [9629 5536] of there. In the Rugby–Harbury area of the Warwick district, the base and top of the member are younger, the former ranging from late *liasicus* to early *angulata* and the latter of late *rotiforme* Subzone age (Old et al., 1987). The non-ammonite fauna includes the bivalves *Cardinia, Gryphaea, Liostrea* sp., *Pinna* sp., and *Plagiostoma giganteum.* Other fossils are listed in George (1937).

Lower Lias Clay Formation

The Lower Lias Clay Formation comprises grey, blocky and fissile, fossiliferous mudstones, with a few thin and nodular limestones. It crops out principally in the south-east of the district, but substantial areas are covered by drift deposits and there are few exposures. The lowermost part of the formation (18.65 m) was proved in the Twyn-ing Borehole, and consists mostly of cyclic alternations of pale grey, bioturbated mudstone and darker grey, laminat-ed mudstone. The spiky nature of the gamma-ray log (Fig-ure 28) suggests that some layers are lime-rich. The Bredon Hill No.2 (Scarborough Cottages) Borehole [9643 3788] (Whittaker, 1972b), sited about 2 km south-east of the district, proved 44.88 m of the upper part of the forma-tion, comprising pale to dark grey, fossiliferous, locally mi-caceous, silty mudstones and siltstones. A siltstone at 40.84 m yielded *Phricodoceras* from near the base of the *jamesoni* Zone and may correlate with the '70' Marker Member of Horton and Poole (1977) which is of early *jamesoni* Zone age. However, this part of the core is faulted and there are no wireline logs. The few limestones mapped occur in the lower part of the formation and are of *oxynotum* Zone age and older. The upper part of the formation becomes silty and passes up into the overlying Dyrham Siltstone Formation. The precise position of the top of the formation around Bredon Hill cannot be mapped as it is covered by landslip deposits.

The formation is estimated to be around 250 m thick, compared to 225 m estimated in the Tewkesbury district (Worssam et al., 1989), perhaps thickening beneath Bredon Hill (Whittaker, 1972b). South-eastwards, it thickens to about 284 m in the Stowell Park Borehole (Green and Melville, 1956), from where it thins on to the London Platform. Thicknesses given by Williams and Whittaker (1974) for the Stratford district are not reliable, the minimum of 61 m quoted for the Ilmington area being probably due to faulting along an extension of the Aston Grove–Preston Fault, although some thinning may have occurred across the Vale of Moreton Anticline. In the Warwick district, the formation thins to between 85 and 100 m.

The age of the formation ranges from the *rotiforme* Subzone of the *bucklandi* Zone to about the base of the *davoei* Zone (Table 12). The Twyning Borehole proved over 20 m of beds of *rotiforme* Subzone age. There are only a few exposures in the district, but the few ammonites recorded represent all the zones up to the *ibex* Zone. The Bredon Hill No.2 (Scarborough Cottages) Borehole proved the *oxynotum*, *raricostatum* and *jamesoni* zones (Whittaker, 1972b). The *ibex* Zone was proved in a slipped block on the south-west flank of Bredon Hill (Worssam et al., 1989).

The beds have a very low, generally eastward dip and are cut by a number of small faults. A major fault bounding the south and south-west of Bredon Hill was proved in the Bredon Hill No.2 Borehole to throw down to the north between 122 and 189 m (Whittaker, 1972b). It could not be traced through the low-lying ground to the north-west of Bredon Hill onto the Worcester district.

Scattered exposures in the more western parts of the outcrop yielded *Arnioceras* indicating a *semicostatum* to *obtusum* zonal age. Spoil dug from a well [c.952 484] north of Pershore Station yielded *Caenisites* and *Angulaticeras* of the *turneri* Zone. An exposure [9289 4757] south-east of Drakes Broughton yielded *Asteroceras* of the *obtusum* Zone and a nearby exposure yielded *Hippopodium ponderosum*. The Atlas Works brick pit, Pershore Station [950 481] exposed "a few feet of blue shaly and slightly calcareous clay with occasional nodules of argillaceous limestone and ironstone" with fossils including "*Ammonites jamesoni*, belemnites, *Gryphaea cymbium*, *G. arcuata* and *Hippopodium ponderosum*" (Woodward, 1893). The record of "*A. jamesoni*" is the only one north of Bredon Hill and from a locality where older rocks are more likely. Ammonites of *obtusum* and *oxynotum* Zone age have been collected here, including *Aegasteroceras*, *Oxynoticeras*, *Promicroceras* and *Slatterites*. An excavation [9519 4806] nearby yielded the *obtusum* Subzone ammonites *Aegasteroceras?*, *Asteroceras* cf. *suevicum*, *Epophioceras* cf. *longicella* and *Xipheroceras* sp.

The low-lying area south and east of Pershore is probably largely of *oxynotum* Zone age. An exposure [9555 4461] at Avonbank yielded *Gagaticeras* of *simpsoni* Subzone (*oxynotum* Zone) age, *Oxynoticeras oxynotum* is recorded nearby [960 451] and the Wick Brickworks [9613 4523] yields specimens of *raricostatum* Zone age. South-west of Avonbank, material from diggings in the river bank at Great Comberton were interpreted to be of *oxynotum* Zone age by Richardson (1905), but the taxa listed do not prove this. The 'brickyard west of Pigeon House' (Wright, 1882, Woodward, 1893) is now infilled [932 476], but formerly exposed 3 m of grey shaly clay with small limestone nodules and an *oxynotum* Zone fauna with *Oxynoticeras oxynotum*. The site also has yielded numbers of *Slatterites slatteri*.

The younger beds of this formation are poorly known, as the slopes of Bredon Hill are extensively landslipped and there are no natural exposures. The non-ammonite fauna includes *Piarorhynchia*, *Ptychomphalus* and bivalves. Faunas are recorded in Richardson (1906), Strickland (1842), Williams and Whittaker (1974), Woodward (1893) and Wright (1882).

Dyrham Siltstone Formation

The Dyrham Siltstone Formation crops out in the southeast of the district on the slopes of Bredon Hill. It is covered by landslip deposits, apart from four small unexposed areas preserved beneath benches formed by the overlying Marlstone Rock Formation.

The formation, as proved in the Bredon Hill No.1 Borehole (Figure 29), consists predominantly of pale to dark grey and greenish grey, micaceous siltstones which are variably muddy, sandy, calcareous and well cemented. They are generally poorly fossiliferous, but with a few shell-debris-rich layers. Burrows, including *Chondrites*, are common. Subordinate silty mudstones, fine-grained sandstones, quartzose wackestones and lime mudstones also occur.

The formation is estimated to be 60 m thick. Over 42.5 m were proved in the Bredon Hill No.1 Borehole, including the whole of the 'Middle Lias Silts' of Whittaker, (1972b), and 7.3 m of the underlying *davoei* Zone siltstones.

The formation spans probably the whole of the *margaritatus* Zone, although the highest part has not been proved, and possibly all of the underlying *davoei* Zone.

Marlstone Rock Formation

The Marlstone Rock Formation, formerly Marlstone Rock Bed, gives rise to a number of prominent spurs on the slopes of Bredon Hill, which are thought to be in-situ (Whittaker, 1972b), and between which the outcrop is covered by landslip deposits. Surface debris and a small outcrop [9517 4059] show grey and brown, ferruginous, fossiliferous limestones and sandy limestones. The Bredon Hill No.1 Borehole (Whittaker, 1972b) proved 6.02 m of pale to dark grey and greenish grey, medium- to fine-grained, fossiliferous sandstones, quartzose wackestones and shell-detrital packstones/grainstones resting on a pebbly sandstone above 50 mm of bored silty lime mudstone. The formation is less ferruginous at Bredon Hill, compared to its outcrop in the Cotswolds.

The formation is poorly exposed and only a limited fauna is known from the district. *Pleuroceras spinatum*, *Tetrarhynchia* cf. *dumbletonensis*, *Pleuromya costata*, *Pseudopecten equivalvis* and belemnites were found in a small quarry [9517 4059] north-north-east of St Catherine's Chapel, Wollashill. Richardson (1902, 1905) recorded *Pleuroceras spinatum* at Wollas Hall [957 406].

Upper Lias Clay Formation

The Upper Lias Clay Formation crops out on the higher slopes of Bredon Hill, but is almost entirely covered by landslip deposits. There are no exposures, but a complete sequence of 110.52 m was proved in the Bredon Hill No.1 Borehole (Figure 29; Whittaker, 1972b). The formation is thought to thicken from about 75 m to 110 m within the district, reaching a maximum to the east, in the centre of the Bredon Hill Syncline (Whittaker, 1972b).

The formation is dominantly silty in the Bredon Hill No.1 Borehole, the lower 21 m comprising medium to dark grey and greenish grey, blocky to fissile and locally fossiliferous, siltstones, silty mudstones, paper shales and scattered limestone nodules. Above are about 80 m of micaceous, fossiliferous, silty to very silty, locally fissile mudstones and calcareous siltstones with rare silty limestone nodules. Pyritic silt burrowfills, some referrable to *Chondrites*, are common. The topmost 8.3 m comprise a coarsening-upward cycle commencing with dark grey mudstones, passing up into 7.4 m of siltstones and sandstones, which are correlated with the Cotteswold Sands of the Cotswolds. They include two thin beds of silty, shell detrital wackestones/packstones, the upper of which contains numerous phosphatic pellets. The top of the formation is placed at 36.04 m, above 6cm of soft, calcareous mudstone and beneath greenish grey, iron-stained sandstones and limestones.

The oldest beds comprise 3 m of blocky siltstones and silty mudstones with micromorphic brachiopods. They may equate with similar beds cropping out about 6 km to the south-east at Alderton and Dumbleton which are of *tenuicostatum* Zone age; this age is not proved here and they are assigned, with the overlying paper shales and thin limestones of typical 'Fish Beds' lithology, to the base of the *falciferum* Zone. The limestones yielded *Cleviceras elegans* and *C.* cf. *exaratum*, proving the *exaratum* Subzone of the *falciferum* Zone. The later *falciferum* Subzone is proved in the siltstones beneath nearly 80 m of silty beds of *bifrons* Zone age, indicating that steady subsidence took place in early Toarcian times.

The whole of the later Toarcian is represented by the topmost 8.3 m of Upper Lias strata and only the latest *aalensis* Subzone of the *levesquei* Zone is proved by the occurrence of *Pleydellia?* in the silty, phosphatic limestone at 36.5 m, just beneath the Inferior Oolite. Neither the *variabilis* or *thouarsense* zones are proved in the borehole, although beds of this age are probably present at Bredon Hill, as ammonites from these zones are known from gravel pits at Overbury and Beckford to the south-east. The grey silty limestone matrix of the specimens from these pits is comparable with the limestone proved between 36.21 and 36.64 m, and this bed may be a con-densed deposit representing much of the later Toarcian. If so, the underlying 89 m of silty beds are the lateral equivalent of the lower part of the Cotteswold Sands and part of the underlying mudstones of the central Cotswold region, north of Stroud, where the Cotteswold Sands are largely of late *bifrons* and *variabilis* Zone age. The highest beds at Bredon Hill would then equate with the Cephalopod Bed of the Stroud district, which is the highest part of the Upper Lias sequence, and indicate a widespread area of slow sedimentation in late Toarcian times. A diverse fauna of brachiopods, gastropods, bivalves, scarce echinoderms, crustacea and fish fragments was also recovered from these beds, which are very poorly exposed at outcrop (Whittaker, 1972b).

INFERIOR OOLITE GROUP

Limestones of the Inferior Oolite Group cap Bredon Hill. They are affected by cambering, but dip generally at 3.5 to 5° towards the south-south-west. The boundary with the underlying Upper Lias is obscured by landslip deposits. The Bredon Hill No.1 Borehole proved 36.04 m, which Whittaker (1972b, Figure 7) considered to be a maximum thickness for the outlier; up to about 22 m are present within the Worcester district.

The lithostratigraphical classification of the group is shown in Table 14. Holl (1863) and Richardson (1902, 1905) recognised four subdivisions of the group on Bredon Hill; in ascending order, these are the Scissum Beds, Lower Limestone, Pea Grit equivalent and Lower Freestone. Later, Richardson (1929) tentatively recognised traces of the overlying Oolite Marl. The stratigraphy of the Bredon Hill No.1 Borehole proved difficult to elucidate owing to the disturbed nature of the strata and the deep weathering effects. Whittaker (1972b) recognised the lowest four units, but did not separate the Lower Limestone and Pea Grit. The uppermost 6.1 m were not cored and the presence of the Oolite Marl was not confirmed. Mudge (1978) reclassified the Inferior Oolite of the Cotswolds and defined several members within the Lower Inferior Oolite Formation. He recognised three members at Bredon Hill, the Leckhampton Limestone, Jackdaw Quarry Oolite and Devil's Chimney Oolite, but Cope et al. (1980b) retained the established names and these are used in this account. A re-examination of the available core samples has allowed the separation of the Pea Grit and Lower Limestone.

The stratigraphy established by Richardson (1902) and Holl (1863) is correct in so far as the order of the beds follows the Cotswold sequence. However the lithologies assigned to two of the beds are at variance with those described here from the Bredon Hill Borehole. The equivalent of Richardson's and Holl's Pea Grit, a compact yellowish limestone with crinoid and echinoid remains, correlates with the Lower Limestone described below. Richardson's Lower Limestone, described as oolitic, is more typical of the Lower Freestone or Pea Grit, and the Lower Limestone in the borehole is poorly oolitic. In the borehole the crinoid-echinoid

Table 14 Classification of the Inferior Oolite of Bredon Hill. Numbers refer to depths in the Bredon Hill No. 1 Borehole [9577 3996]; no depths are given in Mudge (1978).

Whittaker (1972b)	Mudge (1978)	This account
Lower Freestone ——— 13.23	Devil's Chimney	Lower Freestone ——— 13.23
Pea Grit and Lower Limestone ——— 31.16	Jackdaw Quarry Oolite	Pea Grit ——— 25.22
		Lower Limestone ——— 31.16
Scissum Beds 36.04	Leckhampton	Scissum Beds 36.04

limestone facies occurs in the Lower Limestone, Pea Grit and Lower Freestone. These facts suggest that the superficial and tectonic disturbances are more complex than was realised by Richardson and Holl, or that there are some marked lateral facies changes. Both may explain why Richardson's thicknesses differ markedly from those in the borehole.

Fauna from the Inferior Oolite of Bredon Hill is listed in Richardson (1902, 1905, 1929) and Whittaker (1972b). In the borehole the only ammonites recovered were cf. *Ludwigia* and *Bredyia?*, indicating an early Aalenian age.

In the succeeding account, the figures in brackets are the metric equvalents of Richardson's (1902) estimates of thickness from outcrop data.

Scissum Beds

The basal 4.88 m (10.1 m) in the Bredon Hill No. 1 Borehole have been assigned to the Scissum Beds. They consist mostly of pale grey, rust-brown-weathering, fine-grained, silty, slightly micaceous, poorly cemented, quartzose wackestones with lime mudstone nodules and lenses and a little shell debris. Well-cemented sandy, shell detrital wackestones occur near the top and base.

Lower Limestone

The Lower Limestone is 5.94m (10.5m) thick, comprising mostly coarse-grained, shell-detrital packstones/ grainstones, with locally abundant crinoid and echinoid debris. A few ooliths appear towards the top, and the rocks are porous in parts. Subordinate lithologies include shell-detrital wackestones and a thin, poorly cemented sandstone near the top.

Pea Grit

The Pea Grit comprises 11.99 m (4.3 m) of fine- to coarse-grained, generally poorly sorted, moderately well bedded, shell-detrital oolitic grainstones. Bedding is defined by concentrations of shell debris and ooliths and grain size variations. Intraclasts are present in some beds. Subordinate shell detrital wackestone/packstones occur, some being rich in crinoid and echinoid debris. The true pisolitic 'Pea Grit' (Murchison, 1834) of the northern Cotswolds is not present, the rocks all being finer grained.

Lower Freestone

The Lower Freestone consists of fine- to coarse-grained, shell detrital, more or less sandy, cross-laminated, poorly bedded to well-bedded oolitic grainstones. In the specimens examined, cross-lamination is best developed in more sandy grainstones. Subordinate lithologies include crinoid and echinoid-rich packstones/grainstones. The Lower Freestone is 13.23 m (3.7 m+) in the Bredon Hill No.1 Borehole, its base in the borehole at 13.23 m depth marked by a fine-grained, quartzose wackestone/packstone with scattered shell debris.

NINE

Quaternary

The Quaternary deposits of the district range in age from Anglian to Flandrian. Alluvium and river terrace deposits of the Severn, Teme, Avon and Bow Brook make up most of the volume of the deposits. Fluvial, glacigenic and lacustrine deposits infill a palaeovalley to the west of the Malverns (Barclay et al., 1992) and veneers of solifluced, gelifluced and downwashed material are extensive on the flanks of the Malverns.

The stratigraphy of the deposits and the climatic conditions that prevailed during their formation are summarised in Table 15. The oldest known deposit is an interglacial organic silt discovered at the Brays Pit, Mathon [731 443]. The overlying Mathon Sand and Gravel is interpreted as the fluvial deposits of an Anglian river, the Mathon River. These are overlain by the glacio-lacustrine White House Silts, which are in turn overlain by the Coddington Till in the Mathon–Coddington area and by the lacustrine Cradley Silts in the Colwall area. The last probably belongs to the Hoxnian, after which the district remained unglaciated, but cold periods saw the deposition of solifluction and gelifluction deposits.

The river terrace deposits of the district record the climatic variations and events from the end of the Anglian. The River Avon dates from the retreat of Anglian ice from the Severn Valley, although the present-day River Severn came into existence in the late Devensian, when ice blocked its northward flow into the Irish Sea and it was diverted into its present course through the Ironbridge Gorge (Wills, 1924; Worssam et al., 1989).

In addition to the 1:10 000 sheet Technical Reports, details of the Pleistocene deposits of the Malvern Hills area are given by Barclay et al. (1990, 1992).

OLDER FLUVIAL DEPOSITS

Mathon Sand and Gravel

The Mathon Sand and Gravel is the basal deposit of the palaeovalley fill of the Mathon River. It lies beneath the White House Silts and Coddington Till from Mathon southwards. To the north it occurs as dissected remnants, mainly capping small hills and spurs in the Cradley Valley. Three localities have provided good sections through the deposits in recent years: Warner's Farm Pit [7398 4525] (now mostly obscured), Southend New Pit [737 447] (now obscured) and the Brays Pit [729 441]. The last is still being worked at the time of writing and shows the thickest development of sand and gravel (up to 9 m).

Descriptions of the deposits were given by Symonds (1907), Bennett (in Wickham, 1899), Gray (1914), Grindley (1925), Wills (1938) and Hey (1959, 1963). Barclay et al. (1992) gave details of the sections recently or currently available, including clast analysis and sedimentary features.

The deposit is markedly bimodal. Generally, basal gravels are overlain by clean sands. An upper gravel caps the sands locally, as at Warner's Farm. The basal gravel is up to 5 m thick, reddish brown, poorly sorted, horizontally and planar cross-bedded. Clasts are mostly subrounded, typically up to 0.12 m and set in medium to coarse sand. In the 8 to 16 mm fraction, the ratio of local Lower Palaeozoic clasts to exotic clasts, mostly 'Bunter' pebbles and Longmyndian volcaniclastic sandstones, is about equal. Local clasts make up 78 per cent of the 16 to 32 mm fraction. In the cobble fraction, Silurian limestone makes up 50 per cent of the suite, other Lower Palaeozoic lithologies 30 per cent and Malverns Complex lithologies 20 per cent. *Gryphaea* cf. *arcuata* from the Lower Lias is quite common in the larger pebble fraction, as is coal debris aligned along foresets.

The overlying sand is up to 8.5 m thick, mainly red to orange-brown, fine to medium grained, and planar cross-bedded in sets 0.2 to 0.4 m thick. Coal fragments are common as concentrations along foreset laminae, and gravel layers also occur. Scanning electron microscope analysis shows that many grains are aeolian, indicating derivation from the Permo-Triassic aeolian sandstones of the Midlands, but no evidence of glacial textures was recorded.

The upper gravel unit, seen only at Warner's Farm (Plate 13), is about 3 m thick, with a calcreted layer at the top. It is reddish brown, with some darker coaly streaks, flat-bedded and planar cross-bedded. White House Silts (p.88) and Coddington Till (p.86) overlie the gravel.

The deposits are interpreted as those of a braided stream system, typical of British Quaternary cold-stage gravels. However, the lack of glacial textures on the sand grains and striae on the pebbles led Barclay et al. (1992) to preclude a proglacial origin. These authors suggested that the deposits were those of a pre-glacial Anglian river, the Mathon River, and analogous to and synchronous with the Baginton Sand and Gravel of the Midlands and the Shouldham Thorpe Gravels and Sands in Norfolk. The change from gravel to sand was attributed to a reduction in fluvial energy, perhaps triggered by a regional climatic event. The recent discovery of clay clasts in the topmost gravel layers (Dr A Brandon, personal communication, 1992) suggests ponding of the drainage system, most probably by ice as the Anglian glacier reached the catchment area. There is, however, no evidence for the interdigitation of the Mathon Sands and Gravel and the overlying White House Silts, as suggested by Hey (1959), who considered the former to be glacial lake delta deposits.

Table 15 Stratigraphy of the Quaternary deposits of the Worcester district. Correlation with the stages and Oxygen Isotope Stages is tentative.

Stage	Oxygen Isotope	Regional climate	Deposits west of the Malverns	Deposits of the Severn valley	Deposits of the Avon valley
Flandrian	1	Interglacial	Landslip, alluvium, peat, alluvial fan deposits, head		
Devensian	2–4	Stadial/interstadial		First (Power House Terrace)	
		Dimlington Stadial	Younger Head	Second (Worcester) Terrace	First (Birlington) Terrace
	5a	Upton Warren Interstadial		Third (Main) Terrace	Second (Wasperton) Terrace
	5b–d	Stadial/interstadial			
Ipswichian	5e	Interglacial			Third (New Inn) Terrace
Wolstonian	6	Stadial	Colwall Gelifluctate		
	7	Stainton Harcourt Interglacial		Fourth (Kidderminster) Terrace	Fourth (Ailstone) Terrace
	8	Stadial		Fifth (Bushley Green) Terrace	Fifth (Pershore) Terrace
Hoxnian	9 or 11	Interglacial	Cradley Silts	Sixth (Spring Hill) Terrace	
Anglian	10 or 12	Stadial	Coddington Till White House Silts	Glaciofluvial gravel Till	
?Anglian or pre-Anglian		Stadial/interstadial	Mathon Sand and Gravel		
Pre-Anglian	?	Interglacial	Brays Silts		

GLACIGENIC DEPOSITS

Till

The Coddington Till is a red-brown clay with an erratic suite comprising local Malverns Complex and Lower Palaeozoic lithologies and farther-travelled Uriconian, Longmyndian, Carboniferous (including coal fragments) and Permo-Triassic lithologies (Barclay et al., 1992). It occurs in the palaeovalley of the Mathon River and is of variable thickness, its maximum proved thickness being about 10 m at South End [737 447]. It was formerly interpreted as head by Hey (1959; Palmer, 1976), although known to contain coal fragments and striated pebbles. Warner's Farm Pit, Mathon [7384 4538], where it is up to 1.5 m thick, is the most northern known occurrence, although there may be a remnant at Nupend [c.7225 4880]. The till overlies the White House Silts, and, where they are absent, the Mathon Sand and Gravel.

A few isolated occurrences of till are recorded in the Severn valley. Red-brown sandy clays with small indeterminate rock fragments are present on the crest of the Penarth Group escarpment and on the dip slope of the Westbury Formation [924 587] near Netherwood Farm, Oddingley.

Symonds (1883) recorded a section at the Imperial Hotel, now Malvern Girls' College [7835 4585], Malvern, through 'a thick mass of local angular debris, accumulated from the hills above' into a 'stiff red till, or boulder clay, which contained Northern Drift pebbles and angular erratics, rhinoceras bones and mammoth molars and tusks'.

Glaciofluvial Sand and Gravel

Minor occurrences of sand and gravel of possible glaciofluvial origin are attributed to the Anglian glaciation. Isolated remnants of sand and gravel occur north of the River Teme. The largest spread is at Sandy Cross [676 569] and consists of up to 3 m of red-brown sand with scattered small quartz pebbles.

East of the Severn valley, small gravel spreads cap the high ground in the north-east of the district. The largest is at Tibberton [905 570]. It has a markedly inclined base, falling southwards from 63 to 48 m above Ordnance Datum (OD) and consists of up to 2.5 m of reddish brown pebbly sand with pebbles from the Kidderminster Formation. There are exposures of the deposit in the railway cutting at Tibberton.

Plate 13 Warner's Farm Sand and Gravel Pit, Mathon. Observer stands on top of the upper gravel of the Mathon Sand and Gravel; the Colwall Gelifluctate/Coddington Till junction lies immediately above his head.

Elongate spreads of sand and gravel extending south from Churchill Wood [913 541] to Wolverton [913 507] and a deposit north-west of Wolverton [901 516] are probably the remnants of the infill of a channel system. Undulating and channelled bases, variable content of clay and rafts of red mudstone from the Mercia Mudstone suggest subglacial or proximal outwash deposition. There are no sections, but a generalised sequence of the deposits, deduced from a study of soil brash and temporary exposures, comprises clays and sandy clays below pebbly sands, the pebbles being derived from the Kidderminster Formation. The thickness ranges from 1 to 3 m. Richardson (1906) recorded 2.4 m of reddish sand with small quartz and quartzite pebbles in a pit in Churchill Wood. This deposit is close to and at the same level as a high terrace of Bow Brook and has a similar composition. The terrace may be either the Fifth Terrace or a high level of the Fourth Terrace. The former is of late Wolstonian age (Shotton, in Mitchell et al., 1973).

Gravel caps Dripshill [828 457] and consists mainly of 'Bunter' pebbles, with a few subangular Malverns Complex pebbles. It lies at about 74 m above OD and was correlated by Wills (1938) with the Woolridge Terrace. However, the Woolridge Terrace is so variable in height (56 to 122 m) that it is unlikely to constitute a single terrace (Barclay et al., 1992).

LACUSTRINE DEPOSITS

Lacustrine deposits occur in the Cradley valley. All are overlain by other deposits and are either unexposed or ephemerally exposed, before removal, in sand and gravel pits, and thus do not appear on the 1:50 000 map. The White House Silts, of glaciolacustrine origin, were first described by Hey (1959). Two new interglacial lacustrine/channel deposits were discovered during and subsequent to the survey. The Cradley Silts were proved between solifluction or gelifluction deposits (Colwall Gelifluctate) and laminated silts and clays attributed to the White House Silts in an augerhole [7381 4000] near Colwall (Barclay et al., 1992, fig.8). The Brays Silts were found in a small palaeochannel at the Brays Pit [7300 4429].

Brays Silts

The Brays Silts lie beneath the Mathon Sand and Gravel. The palaeochannel in which they occur has been intermittently exposed since 1990. Its fill comprises grey, organic silts rich in an interglacial flora and fauna of conifer branches and cones, and insect remains (Dr A Brandon, personal communication, 1990).

White House Silts

The White House Silts lie between the Mathon Sand and Gravel below and the Coddington Till above. They are reddish brown, grey and yellow silts and clays and fine sands, commonly finely laminated and probably varved. Autochthonous carbonate concretions are common. Hey (1959, 1963) applied the name to a sequence of laminated clays, silts and fine sands found in the Cradley valley between Mathon and the area of White House Farm [7265 3844], near Eastnor in the Tewkesbury district to the south. Up to 7 m are present in the Mathon area, but thicknesses are variable. Only 0.2 m are locally present at Warner's Farm Pit, the northernmost proved occurrence, where they comprise folded reddish brown and yellow silts. A small patch of silt at Nupend [7255 4880] in the north of the Cradley valley overlies the Mathon Sand and Gravel, and may correlate with the White House Silt.

Pinkish brown and reddish brown laminated silt proved at the base of an augerhole [7381 4000] near Colwall (Beds A1 and A2 in Barclay et al., 1992, fig.8) is correlated with the formation. It is devoid of Quaternary pollen, apart from one grain of *Cyperaceae*.

Cradley Silts

The Cradley Silts are fossiliferous grey silts, discovered by augering through gelifluction deposits on the broad flat watershed area between the Cradley and Glynch brooks near Colwall. Two augerholes [7381 4000] at the same site provide the only record (Barclay et al., 1992, Beds 3, 4, 5 and 6, fig.8). The silts rest on White House Silts, the Coddington Till being absent.

The silts are up to 1 m thick, and contain a thin peat near their top. Sand grains, molluscs and ostracods occur throughout and small rock fragments occur at the base. Pollen analysis of the peat, and details of the fauna are given in Barclay et al. (1992). The pollen profile suggests an age straddling the Anglian–Hoxnian boundary, the faunas indicating deposition in a body of stagnant to slow-moving water in a temperate climate.

FLUVIAL DEPOSITS

In recent years there has been a resurgence of interest in the river terraces of the Severn and Avon and their tributaries. The recent work has built upon the framework established by Tomlinson (1925) for the Avon, and Wills (1938) for the Severn terraces, and has indicated (Whitehead, 1989a, b; Dawson, 1989) that climatic vacillations and sedimentation during the development of the two river systems were more complex than previously assumed. Recent advances in amino acid geochronology have enabled the terrace sands and gravels to be dated and compared with deposits of similar age elsewhere in the British Isles. These results have necessitated several changes to the previously accepted correlations (Shotton, 1973) of Quaternary deposits in the British Isles.

Tomlinson (1925) recognised and numbered five terraces of the River Avon above the level of the present flood plain. She considered that the Third and Fourth terraces resulted from a single aggradation to the higher (Fourth Terrace) level, followed by later downcutting to the lower (Third Terrace) level, thus making the deposits underlying the Third Terrace feature older than those underlying the Fourth. Her conclusions were supported by Shotton (1968), but later rejected by Whitehead (1989a) and Bridgland et al. (1989), who considered that the available evidence indicates two separate periods of aggradation. Maddy et al. (1991) proposed a new nomenclature for the deposits of the Avon valley, introducing several formally named members, equivalent to the numbered terraces proposed by Tomlinson (1925). Recent mapping by the BGS has shown that locally both the Second and First Terrace deposits can be subdivided into two units.

In the Severn valley, Wills (1938) recognised and named five terraces. A sixth, higher terrace, grouped with the Fifth Terrace by Tomlinson and Wills, has been recognised by the BGS (Figure 31; Worssam et al., 1989) in the Severn and Avon valleys. Maddy et al. (1995) have erected a nomenclature for the Severn valley deposits, similar to the earlier scheme for the Avon valley, giving member status to the terrace deposits, and naming the suite the Severn Valley Formation.

A simple correlation between the terraces of the Avon and Severn suites cannot be made (Table 15). For example, terrace deposits equivalent to the Third Avon Terrace have not been recognised in the Severn valley, and the Third Terrace of the Severn appears to broadly correspond to the Second Avon Terrace. Five terraces of the River Teme and its tributary Sapey Brook have been identified within the district. The Second and Third terraces are equivalent to the Second (Worcester) and Third (Main) terraces of the Severn respectively (Wills, 1938). The equivalence of the First, Fourth and Fifth terraces is less certain.

The River Avon came into existence following the retreat of ice in the Severn valley. This glaciation was considered to be of Wolstonian age by Shotton et al. (1977), although others (Bowen et al., 1986; Sumbler, 1983a, b; Rose, 1987) believe it to be Anglian, and therefore of equivalent age to the glaciation which produced the chalky till covering much of eastern England. Prior to this glaciation a large part of the Midlands, from about Bredon Hill north-eastwards, may have been drained northwards through Leicestershire into the River Trent via the River Soar. Following the retreat of the ice the Avon may have provided the main drainage of the area, with the River Severn, a tributary of the Avon, rising near Kidderminster. This continued until between 25 000 and

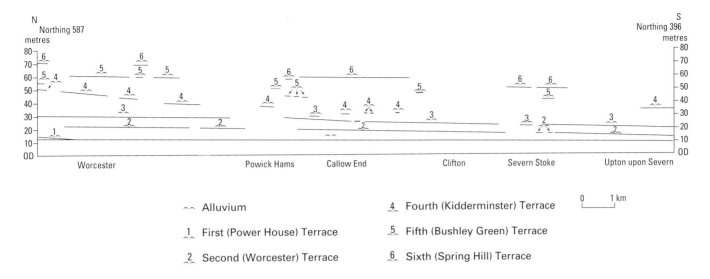

Figure 31 Surface levels of the river terraces of the Severn valley.

13 000 years ago (middle to late Devensian). At that time an advancing ice sheet blocked the northward course of the upper part of the present day Severn, which until then had flowed into the Irish Sea. The river was diverted into its present course, cutting the Ironbridge Gorge, north of Bridgnorth. Since then, the Severn has been the major river, with the Avon a tributary. The absence in the Severn valley of a terrace equivalent to the third terrace of the Avon may be due to erosion of these deposits following the cutting of the Ironbridge Gorge.

The Severn terraces are composed predominantly of clasts of well-rounded 'Bunter' quartz and quartzite, commonly liver coloured, derived from the Triassic Kidderminster Formation. Angular and subangular igneous clasts derived from the nearby Precambrian Malverns Complex of the Malvern Hills to the west are also present, particularly along the west side of the Severn valley. Other pebbles include flint, Triassic sandstones, siltstone, mudstone, and coal. In addition, Wills (1938) recorded pebbles of rocks from Scotland, the Lake District, Wales and the Welsh Borderland. All these were derived from glacial drift in the Midlands and the Welsh Borders, the Welsh material being restricted to the lower terraces deposited after the cutting of the Ironbridge Gorge.

The main constituents of the Avon terraces are rounded clasts of 'Bunter' quartzite and vein quartz, Jurassic limestone pebbles and flints derived from the Chalk (Tomlinson, 1925).

In several places it has not been possible to allocate spreads of gravels to a specific terrace. In such instances the gravels are shown on the map as 'River Terrace Deposits, undifferentiated'.

Terrace Deposits of the River Severn and River Teme

SIXTH (SPRING HILL) TERRACE

Wills (1938) recorded thick accumulations of gravels in the Severn valley and assigned them to a single period of aggradation, naming them the Bushley Green or Fifth Terrace. Recent mapping by BGS in the Tewkesbury (Worssam et al., 1989) and Worcester districts has revealed the presence of two distinct deposits, locally separated by a bedrock step, within the height range of Wills' Bushley Green Terrace. The higher terrace was assigned to a Sixth Terrace by the BGS surveyors and later named the Spring Hill Member by Maddy et al. (1995).

The base of the Sixth Terrace lies at about 65 m above OD in the north, declining southwards to about 50 m above OD at Severn Stoke [85 44], some 50 to 40 m respectively above the present floodplain. Field observations and resistivity soundings indicate that the deposits have a maximum thickness of about 4 m throughout the district.

To the west of the Severn, deposits are restricted to several small patches near Upper Broadheath [805 558] and Callow End [831 490]. East of the Severn, two small patches of gravel are present at Hindlip [880 585] and a large irregular spread caps the high ground at Stonehall [880 490], where the deposit consists mainly of red to orange-brown sand, seen in disused workings on Kempsey Common [871 481]. Two spreads are present [865 443; 860 430] to the west of Kinnersley.

Gravels also occur at the Sixth Terrace level near Strensham, about 35 to 40 m above the Avon floodplain. These deposits, although broadly within the present Avon catchment, are considered to be part of the Severn terrace sequence, as there is no evidence higher up the Avon valley for any Sixth Terrace deposits. At the surface 'Bunter' quartzite and flint pebbles are common in fields, and augering shows the gravel to have a clayey matrix. Tomlinson (1925) included these deposits in the Fifth Terrace as she considered them to be part of a continuous spread of gravels stretching down to the Fourth Terrace deposits near Bredon Field Farm [90 39]. However, there is a distinct feature marking the base of the

Sixth Terrace deposits, below which Lower Lias clays form a gentle slope down to the Fifth Terrace 5 m below.

FIFTH (BUSHLEY GREEN) TERRACE

The Fifth Terrace was named the Bushley Green Terrace by Wills (1938) and the Bushley Green Member by Maddy et al. (1993).

The base of the terrace deposits ranges from between 50 and 60 m above OD in the north to about 40 m above OD in the south, lying 35 to 28 m above the present floodplain.

Outcrops west of the River Severn are restricted in this district to an extensive patch at Lower Broadheath [810 570], some smaller areas around Upper Broadheath to the north of the confluence with the Teme, and to several small patches south of the Teme. East of the river there is a small patch north of Worcester [860 587] and another about 1 km east [869 472] of Kerswell Green, consisting of 1 to 2 m of pebbly sand.

Sections at the type locality [8620 3508] on the west side of the valley just south of the district have revealed a varied fauna including *Mammuthus primigenius* and *Rhinoceros* (Lucy, 1872). A shelly fauna, first identified by Kennard (Wills, 1938, p.176), and described by Bridgland et al. (1986), indicates a climate similar to that of today. The fauna is in the basal part of the sequence, which consists of horizontally bedded, muddy sands and coarse, gravelly sands. Above are planar cross-stratified, fine sands. A distinctive solifluctied layer of coarse limestone and flint boulders lies within the upper sands. A top layer comprises unbedded, cryoturbated, red-brown clayey sands, with gravel lenses at the base. This was interpreted by Wills (1938) as a till, but regarded by Bridgland et al., as a slope deposit. Amino acid epimerisation analyses of *Trichia hispida* have yielded a mean ratio of 0.235 ± 0.01, correlated by Bowen et al. (1989) with the temperate deposits at Hoxne, Suffolk, the type site of the Hoxnian Stage and Oxygen Isotope Stage 9. Maddy et al.(1995) suggest that the Bushley Green deposits are transitional over the Stage 9/8 boundary, with Stage 8 being represented by the cold stage sands of the middle unit.

A small patch of gravel rich in locally derived calcrete clasts at Whitbourne Hall [704 568], with a base level just over 70 m above OD is tentatively correlated with the Fifth Terrace.

FOURTH (KIDDERMINSTER) TERRACE

In the Worcester district the Kidderminster Terrace is confined to a number of dissected outcrops of gravel to the west of the Severn and some small patches of pebbly sand east of the river near Kempsey [864 494] and at Napleton [862 484]. The base of the terrace falls from 45 m above OD in the north to 30 m above OD in the south, 30 m and 18 m respectively above the present floodplain level.

Wills (1938) mapped the deposits from Kidderminster to Bewdley, and described them as sands and gravels, the latter comprising mainly 'Bunter' quartzites and vein quartz, with an absence of Irish Sea erratics.

Deposits tentatively correlated with this terrace occur in the Teme valley at Lower Poswick Farm [7090 5715], where up to about 1.5 m of gravel are present, and east of Folly Farm [756 451]. The base of the former is between 60 and 65 m above OD, that of the latter at about 55 m above OD.

THIRD (MAIN) TERRACE

The Main Terrace, the Holt Heath Member of Maddy et al., 1995, is the most extensive of the Severn terraces in the district, occurring in outcrops several kilometres long, on both sides of the river. Smaller outcrops are also present along the northern side of the Teme valley.

Sections available now and in the past indicate that, in general, the deposits comprise gravels overlain by sand. The gravels contain a high proportion of exotic material of Scottish and Lake District origin, similar to clasts in the Devensian till farther north. Near Hanley Castle, the gravels become enriched in angular clasts of Malverns Complex granites and diorites and merge westwards into sheets of poorly sorted gravels which extend to the base of the Malvern Hills some 6 km distant.

In the north the base of the deposits lies at 25 m above OD, 10 m above the present floodplain, dropping southwards to 15 m above OD, 5 m above the floodplain. The base is locally irregular and incised into the underlying bedrock by up to 3 m, for example 300 m north-west [856 461] of Naunton Farm, Birch Green. Conductivity meter (EM31) traverses (Ambrose et al., 1987) indicate that, in places, the base of the Main Terrace lies at a lower level than the upper surface of the Second (Worcester) Terrace, a relationship not recognised by Wills (1938).

Extensive spreads of gravel occur in the Teme valley around Cotheridge Court [785 546], at least 2.5 m of gravel being reported east of Cotheridge [7990 5457].

Details of fossils and artefacts found in pits in the Third Terrace were given by Wills (1938, pp.191, 192). These are curated in Worcester Museum, and include a molar of *Mammuthus primigenius* from a 'pit near Powick farm house' and a similar molar, labelled 'Bromwich Hill', probably from a disused quarry near Pitmaston House [c.838 538], Vernon Park. Artefacts include Acheulian flint implements and a Mousterian flake. Marine shells are considered by Wills to have been derived from Irish Sea drift.

Hippopotamus remains were obtained by Boulton (1917) from the base of gravels believed to correlate with the Main Terrace at Stourbridge in the Stour valley, a tributary valley of the River Severn. Hippopotamus is generally regarded as an indicator of the Ipswichian Stage (Stuart, 1982), which correlates with Oxygen Isotope Sub-stage 5e. Wills (1938) considered these specimens to have been reworked from an older deposit, but it is possible that deposition of the Main Terrace commenced during Sub-stage 5e (Maddy et al., 1995).

There are also organic deposits in the base of this terrace at Upton Warren in the River Salwarpe valley, a tributary of the River Severn. This is the type locality of the Upton Warren Interstadial (Coope, Shotton and Strachan, 1961), which, until recently, was assigned a mid-Devensian age. However, amino acid epimerisation determinations indicate a slightly older date which correlates with Oxygen Isotope Sub-stage 5a (Bowen et al., 1989).

Wills (1938) described gravels of the Main Terrace interbedded with glacial deposits at Eardington, Birmingham. Maddy et al. (1995) consider that the bulk of the Main Terrace deposits originated as outwash from a Devensian ice sheet.

SECOND (WORCESTER) TERRACE

The Worcester Terrace forms extensive areas contiguous with the modern floodplain along both sides of the Severn. Resistivity soundings demonstrate that the gravels extend under the alluvial deposits of the modern floodplain (Ambrose et al., 1987). The upper part of the deposits is typically more sandy than the lower; winnowing at the surface concentrates the gravel fraction of the upper part and may give a misleading impression of the gravel content. The deposits are between 5.5 and 10.4 m thick in the Draycott–Severn Stoke area. Their base is uneven, ranging from 12 to 21 m above OD in the north and falling to 8 m above OD in the south, 3 m below the present floodplain. Contours on the base of the terrace in the Worcester area indicate the presence of north–south trending channels (Morris, 1974).

In the Teme valley there are extensive deposits near Cotheridge [79 54] and north of Knightwick [728 572; 732 580], and smaller patches around Bransford [805 527]. A section [734 583] at Horsham showed about 2.3 m of red silt above 1 m of gravel.

FIRST (POWER STATION) TERRACE

To the north of Worcester the Power Station Terrace is, within the district, restricted to several small patches 1 to 2 m above the level of the present floodplain. Downstream of Worcester the terrace deposits fall below the level of the modern alluvium. Resistivity soundings indicate that up to 6.6 m of sand and gravel are present. To the north-west, a terrace lying 1 to 2 m above the floodplain of the river in the Teme valley is tentatively correlated with the First Terrace.

Terrace Deposits of the River Avon

FIFTH (PERSHORE) TERRACE

Gravels assigned to the Pershore Terrace occur at Strensham [910 400] about 37 to 42 m above the Avon floodplain, at Besford [915 445] and at Tyddesley Wood [929 488]. Sloping terraces [928 415] at Eckington, with a base ranging from 35 to 41 m above OD, were correlated with the Fifth Terrace by Tomlinson (1925) and Wills (1938); deposits to the south [930 407], with a height range of 35 to 45 m above OD, were correlated by Wills with the Fourth Terrace. Both deposits are here assigned to the Pershore Terrace.

The disarticulated skeleton of a mammoth was discovered in 1990 at Upper Strensham [904 392], just to the south of the present district (De Rouffignac et al., 1995). The skeleton was in clays beneath gravels mapped as

Fifth Terrace deposits. Associated flora and fauna indicate deposition of the clays in an abandoned river meander in marshland surrounded by an area of restricted tree cover. Amino acid geochronology suggests that the deposits represent the Oxygen Isotope Stage 7 interglacial period. However, preliminary amino acid determinations from Fourth (Ailstone) Terrace deposits at Ailstone (see below), to the north of the district, also indicate a Stage 7 age (Bridgland et al., 1989). Further work is required to resolve this matter.

At Pershore, a section [038 464] seen in 1979 revealed a channel cut in Lower Lias clay (Whitehead, 1989a) and filled with sand and gravel. The base of the channel lies 39 m above the present floodplain. Although these deposits are believed to correlate with the Pershore Terrace, Whitehead (1989a) urges caution, suggesting that they may be older, having been buried beneath Pershore Terrace deposits which were subsequently eroded. A flora from organic material contains the deep-water hydrophytes *Groenlandia densa* and *Potamogeton praelongus* and the ostracod *Limnocythere sanctipatricii*, which prefers deeper sluggish water. The bivalve *Pisidium moitessierianum* and bones of Red Deer (*Cervus elaphas*), coupled with the evidence from the flora, indicate a temperate climate similar to the present.

FOURTH (AILSTONE) TERRACE

Within the Avon catchment area extensive remnants of the Ailstone Terrace are preserved along the Bow Brook, a tributary of the Avon, suggesting that at that time the brook was a major river.

Deposits of the Ailstone Terrace at Birlingham [928 438] and Home Farm [912 446] have their base at 38 to 44 m above OD, 24 to 30 m above the Avon floodplain. Tomlinson (1925) recorded Fourth Terrace deposits around Birlingham Court [9325 4359], but only thin gravelly wash resting on Lias clay was augered during the survey. It is improbable, therefore, that the well at Birlingham, mentioned by Tomlinson as having proved at least 6.1 m of gravel with freshwater snails, was situated at Birlingham Court.

Evidence of climatic conditions during the deposition of the Ailstone Terrace deposits is contradictory. At Twyning, just south of the district, the deposits comprise two members (Whitehead, 1989a). The lower member has a full glacial biota comprising *Mammuthus primigenius*, *Coelodonta antiquitatis*, *Equus* cf. *spelaeus*, *Rangifer tarandus*, *Bison*, and the mollusc *Pisidium vincetianum*. The upper member contains ice-wedge pseudomorphs and yielded a fresh valve of *Corbicula fluminalis* in-situ; the former indicates cold, periglacial conditions, the latter interglacial conditions. The aggradation of the terrace appears to have taken place during a period of climatic change from glacial to interglacial conditions (Whitehead, 1989a).

At Ailstone, the Fourth Terrace deposits have been reinvestigated by Bridgland et al. (1989), who record a fully interglacial molluscan fauna. Preliminary amino acid determinations (Bridgland et al., 1989; Bowen et al., 1989) suggest that the deposits equate with those at Stanton Harcourt, which are attributed to a currently undefined interglacial between the Hoxnian and the Ipswichian, thought to correspond with Oxygen Isotope Stage 7 (Shotton, 1983).

THIRD (NEW INN) TERRACE

Deposits assigned to this terrace occur at Defford, Birlingham, and along the sides of the Bow Brook, generally sloping up from the contiguous and flatter Second (Wasperton) Terrace. The base of the deposits lies between 25 and 30 m above OD, 10 to 15 m above the present floodplain. In addition to 'Bunter' pebbles and flints, some oolitic limestone pebbles are present.

A small remnant of terrace forms the core to a meander of the Avon at Eckington, where Strickland (1842) recorded mammalian bones, found mainly at the base of the deposits in a railway cutting [c.920 415]. These include *Mammuthus primigenius* (possibly *Palaeoloxodon antiquus* according to Tomlinson, 1925), *Hippopotamus major*, *Bos* and *Magaceros*. Numerous freshwater shells were present, but Strickland named only *Pisidium amnicum* (*Cyclas amnica*) and *Sphaerium corneum* (*Cyclas cornea*). An extensive interglacial molluscan fauna, together with *H. amphibius* and *Bos* sp., was found nearby [919 417] by Keen and Bridgland (1986), who concluded that the fauna is of Ipswichian age. This throws some doubt on the identification of *Hippopotamus major* by Strickland as this species occurs in the Lower Quaternary, whereas *H. amphibius* is characteristic of the British Ipswichian (Stuart, 1982). Keen and Bridgland recorded 39 molluscan taxa, of which 21 are freshwater and the remainder terrestrial. The dominant freshwater gastropods, *Valvata piscinalis*, *Bithynia tentaculata* and *Ancylus fluviatilis* indicate fluvial deposition. The others are known to live in clear, fresh ponds. However, bivalves such as *Pisidium henslowanum* and *P. moitessierianum*, which are usually associated with large rivers, are present only in small numbers. The bivalve population is dominated by *Sphaerium corneum* and *Psidium casertanum* which are usually pond species. Keen and Bridgland concluded that the gastropods provide the more reliable environmental indicator and that the deposit is fluvial in origin. The terrestrial fauna is dominated by grass and marshland taxa. The occurrence of the shade-loving *Discus rotundatus* and *Clausia* sp. suggests the presence of shaded habitats of scrub or woodland away from the river floodplain (Bridgland et al., 1989).

Bridgland et al. (1989) concluded that these deposits are younger than the Ailstone Terrace deposits (cf. Tomlinson, 1925).

SECOND (WASPERTON) TERRACE

Deposits of the Wasperton Terrace form extensive spreads within several broad meander cores between Pershore [94 45] and Eckington [99 41], with smaller spreads along the valley of the Bow Brook. The base of the deposits lies about 5 to 8 m above the present floodplain. On the west side of the Avon, between Defford [915 430] and Bourne Brook [905 415] the terrace has been divided into two levels (2a and 2b) separated by a low rise. There is insufficient evidence to indicate whether these levels correspond to benches in the underlying Lias bedrock, as near Twyning in the district to the south. The Second and First Terrace deposits are generally contiguous; where the latter is absent, the Second Terrace deposits are separated from the modern alluvium by a rock step, with the base of the terrace commonly marked by a spring line.

In addition to abundant 'Bunter' quartzite pebbles and clasts of angular flints, the deposits contain rolled Jurassic fossils and pebbles of Jurassic limestones, particularly on the south bank of the Avon. Limonite pellets are abundant in the finer fraction.

From the railway cutting at Defford [c.915 428] Strickland (1842) recorded *Mammuthus primigenius*, *Coelodonta antiquitatis*, *Hyaena* and freshwater shells.

FIRST (BIRLINGTON) TERRACE

The most extensive deposits of the First Terrace are preserved in the cores of three meanders between Pershore [94 45] and Birlingham [93 42]. Near Birlingham and Pensham [93 44] they are divided into a lower (1a) and upper level (1b) based on a slight variation in surface height. Other smaller outcrops have beem mapped along the course of the Bow Brook, which enters the Avon to the south-west of Pershore.

The deposits are contiguous with the modern alluvium, with a slight rise of about 1 m at the junction between the two deposits. The upper surface of the terrace rises gently to about 3 m above the alluvium at the back edge of the terrace.

The deposits are predominantly of silt, sand and gravel, but a shallow linear depression, about 150 m wide, west of Pensham [938 445] is underlain by clayey soils and probably represents an ancient cutoff meander loop.

A limited fauna has been obtained from the deposits. A radiocarbon date of 9030 ± 200 BP (Shotton, 1973) from deposits overlying the main gravel may date from a later, exceptional high flood event.

Alluvium

The main rivers and their tributaries are flanked by alluvium up to several metres thick and in some instances forming belts more than one kilometre wide. The general sequence proved in boreholes beneath the Severn, Avon and Teme alluvial plains is of silt and clay overlying sand and gravel. Much of the sand and gravel is probably Devensian, although some may be Flandrian in age. In the smaller valleys the basal sands and gravels are typically absent.

The alluvial clays and silts of the Severn valley are predominantly red-brown to brown, reflecting the colour of the Mercia Mudstone bedrock from which the deposits are largely derived. Locally, the deposits are sandy or contain organic debris. Within the Avon catchment the alluvial deposits are typically greyish brown, reflecting the colour of the Jurassic source rocks.

The broad tract of alluvium at Smithmoor Common [SO 87 41] east of Upton upon Severn comprises dark grey to brown clay which grades locally into red-brown clay towards the margins. This deposit appears to be too extensive to have been deposited by the present drainage, and may have been laid down in a temporary lake.

Alluvial Fan Deposits

Alluvial fans occur at the break of slope where tributary streams deposit their bedload before joining the main rivers. The larger fans in the district occur along the Teme Valley. There are few sections, but the deposits are likely to be similar to the alluvium of the main rivers.

ORGANIC DEPOSITS

Peat

In addition to the peat within the Cradley Silts (p.88), peats of Flandrian age occur to a limited extent, resting on alluvium. A deposit at Moor End [911 561] consists of at least 1.2 m of peat and intercalated organic clay. An area of clayey peat and organic clay [899 521], at least 1.2 m thick, is present east of Old House Farm. At least 1 m of peaty clay has been proved in the valley of Hatfield Brook [865 504] near Holdings Farm. Peat occurs in a narrow strip at Ashmoor Common [852 466] in the valley of an unnamed tributary of the River Severn between Clifton and Kerswell Green. It is generally less than 1 m thick, grading into organic clay. The oldest peat at this locality has been dated as 5930 ± 70 BP (Brown, 1983).

SLOPE DEPOSITS

Head

Grouped under the term Head on the 1:50 000 map is a range of deposits of solifluction, gelifluction and colluvial origin. Only the Colwall Geliﬂuctate is named; Older Head is distinguished in the Cradley valley and some deposits on the flanks of Bredon Hill in the south-east of the district are correlated with the Avon terraces into which they grade.

The Colwall Geliﬂuctate forms a dissected terrace in the Cradley valley and floors the broad watershed area near Colwall. It generally comprises up to 2 m of yellow-brown gravelly silty clay and clayey silt with abundant angular clasts of Silurian and Malverns Complex lithologies, many of them arranged with their long axes vertical due to cryoturbation. The type section for the deposit is in augerholes at Colwall [7381 4000], where it is 4.3 m thick and consists of two stony units separated by a laminated clay with few stones (Barclay et al., 1992, fig.8, Beds 7 to 9).

Older Head, termed Geliﬂuctate (Younger) by Barclay et al. (1992), is differentiated in the Cradley valley, where it floors dry valleys. It consists of up to 2.5m of yellowish brown to greyish brown, clayey, sandy silt with angular clasts of Malverns Complex and Silurian lithologies, mainly concentrated in the lower part.

The most extensive deposits shown as Head (undifferentiated) on the 1:50 000 map occur on the flanks of the Malverns. The deposits on the east side of the hills comprise gravelly silts and clays, with angular Malverns Complex clasts in red clay derived from the underlying Mercia Mudstone Group. In the Tewkesbury district to the south, these deposits are termed Fan Gravels and several generations are distinguished on the basis of the terraces of the Severn into which they grade. The name Fan Gravel is no longer used in the current BGS drift classification, and, in this district, the deposits form a continuous apron rather than fans. Also, they are rarely gravelly enough to be termed gravels. The deposits generally range up to 2 m in thickness, but their thickness is variable, as they veneer an irregular rockhead topography. The thickest recorded occurrence is at Malvern Girls' College [783 4585], near the site of Symonds' (1883) till record (p.86), where boreholes show up to 6.5 m of stony, sandy clays with gravel pockets. In general, the deposits are mapped where they are thicker than 1 m, but a thinner veneer can be expected in areas shown as drift-free on the 1:50 000 map from Malvern northwards to the River Teme.

Several patches of head have been mapped to the north of the Malverns between Leigh Sinton and Bransford [794 510]. These form near-flat terrain, and consist of pebbles and cobbles of Malverns Complex and Llandovery sandstone, with some Silurian limestone and 'Bunter' quartzite. The clasts are mainly subangular, apart from the 'Bunter' pebbles, and typically 0.01 to 0.15 m in length. They are poorly sorted and set in a matrix of clay, silt and sand. The 'Bunter' pebbles are derived from river terrace deposits.

On the western and northern flanks of the Malverns, head deposits up to 3 m thick consist of yellow to grey silts and clays clasts of Malverns Complex and Silurian lithologies. Away from the Malverns, head is ubiquitous, but is generally only mapped in its thicker occurrences in valley bottoms. East of the River Severn, the most widespread mapped occurrences are south-east of Worcester [885 510] where stony, gravelly, sandy clays, derived from Sixth Terrace deposits, are up to 2 m thick.

The deposits on the flanks of Bredon Hill have been described as Fan Gravels (Whittaker, 1972b; Old, 1987). Near Clattsmore Farm [932 400] roughly bedded limestone gravel slopes from 45 to 33 m above OD towards the River Avon. A small patch of similar gravel nearby [9365 4026] slopes westwards from 70 to 60 m above OD. These deposits are similar to those described by Tomlinson (1941), who attributed them to deposition by seasonal torrents of meltwater in a periglacial environment. The lower fan appears to grade into the level of the Avon Fourth Terrace, with which it is correlated. Apparently unstratified limestone gravel with a clay matrix forms a fan sloping northwards from Woollas Hall Farm [945 410] to Nafford and another, smaller one occurs to the east. These grade down to the level of the Avon Second Terrace and were interpreted by Tomlinson (1941) as taele-fan gravels produced by gelifluction.

Landslip

Extensive landslip deposits are confined to the flanks of Bredon Hill in the extreme south-east of the district, and to the flanks of Ankerdine Hill in the north-west. There are also large areas of landslip on the Lias outcrop on the banks of the Avon.

Landslips in the undulating Old Red Sandstone country in the west of the district are mostly shallow earthflows of clayey material weakened by water emanating from spring lines at the bases of sandstones or calcretes.

On the Silurian outcrop, the thinly bedded sandstones, mudstones and bentonitic clays of the Wyche Formation are responsible for many shallow earthflows. The most extensive area of landslips is around the flanks of Ankerdine Hill, where the steeply dipping beds on the limbs of the Malvern Axis contribute to the instability. Hollingworth (1938) noted active landslips; fresh scars and earthflow lobes testify to the continuing instability of this area.

To the north of Ankerdine Hill, earthflow deposits are present on the eastern flanks of the Teme Valley, derived from the outcrops of the Wyche, Raglan Mudstone and Highley formations. To the south, a large area of shallow earthflows is present in the Lord's Wood area [735 552], involving clays derived from the Silurian outcrop, including the Raglan Mudstone Formation.

Amongst several areas of shallow earthflow failure in the Storridge–West Malvern area is a rotational failure in the Woolhope Limestone in High Wood [7585 4735]. A small landslip was induced at Cowleigh Bank [7670 4745] in 1979 during excavations for a service reservoir, causing the abandonment of the project. The slip took place on a steep, clay head-covered slope underlain by weak Silurian, Triassic and Malverns Complex lithologies where the East Malvern Fault and Precambrian–Llandovery junction converge.

East of the Severn, there are two areas of landslip near Upton Snodsbury, one [940 550] in the Rugby Limestone, and the other [937 551] mostly within the underlying Saltford Shale. Both are shallow earthflows, the latter induced by groundwater emanating from the base of the limestone. There are several areas of hummocky to stepped rotational landslips on the banks of the River Avon from Pershore to Strensham where Lower Lias clays form steep banks. These are at Pershore [947 447], Birlingham [93 43], Strensham [91 40] and from Nafford [94 41] to Great Comberton [95 42]. Active movement was noted in the landslips at Great Comberton and minor failure of the A4104 was noted in 1985 at Tyddesley Wood [9306 4445] on the landslip at Birlingham.

A wide zone of landslip deposits occupies the northern and western slopes of Bredon Hill, extending from the base of the Lower Inferior Oolite to below the base of the Dyrham Siltstone Formation. The landslips, described by Whittaker (1972b), include major rotational slips, especially above the outcrop of the Marlstone Rock Formation, earthflows and mudflows. A currently active mudflow above St Catherine's (Woollas Hill) Farm [9435 4001] was described by Grove (1953) and Whittaker (1972b). Whether landslipping alone is responsible for the 15 to 20 m discrepancy in height of the Marlstone Rock Formation on adjacent spurs in this vicinity, as suggested by Whittaker (1972b), is unclear from present evidence. Faulting may also be present, and a northwest–trending fault is shown on the 1:50 000 map.

MADE GROUND

Made ground deposits of the district are subdivided into two categories: those that lie on the original ground surface and those backfilling or partially backfilling pits, quarries and excavations. Since much of the district is rural, made ground is volumetrically insignificant, most of the larger areas being concentrated in and around the city of Worcester. In addition, much of the urban area is covered by a thin layer of made ground not shown on the 1:50 000 map. The backfilled pits in the Severn valley are mostly former gravel pits, with some clay pits. Probably the thickest occurrence of fill is in a former brick pit at Malvern [771 476] (Plate 8), where up to 26 m of domestic refuse were proved in boreholes.

TEN

Structure

The district consists of two distinct structural areas; the Lower Palaeozoic outcrop in the west and the Worcester Basin in the east (Figure 32). Separating these is the north–south-trending Malvern Axis which marks a Proterozoic zone of crustal weakness that has undergone repeated reactivation. In the south of the district, the Precambrian, igneous Malverns Complex, together with a small amount of volcanic rocks of the Warren House Formation, crops out on the axis in the Malvern Hills. To the north, the axis comprises a narrow zone of anticlinally folded and locally overthrust Silurian rocks. Figure 33 shows a regional cross-section, close to the northern margin of the district, based upon seismic reflection, borehole and outcrop data. To the west of the Malvern Axis, a relatively flat-lying Precambrian basement is overlain unconformably by a cover of Lower Palaeozoic rocks in which the dominant structure is a suite of north-west-trending faults and folds. To the east of the Malvern Axis is the Worcester Basin, a Permo-Triassic extensional graben, formed after Variscan uplift and erosion of the Lower Palaeozoic cover. The western boundary of the basin is the north-trending East Malvern Fault. The eastern boundary of the basin, the Inkberrow Fault, lies just to the east of the district.

THE MALVERN AXIS

The Precambrian deformation history of the igneous Malverns Complex is described in Chapter 2. The structure of the Malvern Axis has been the subject of much debate (e.g. Penn and French, 1971, Bullard, 1989). It has been interpreted as a faulted nappe (Raw, 1952), an upthrust block (Groom, 1899, 1900, 1910; Blyth, 1953; Phipps and Reeve, 1969) and a thrust-faulted monocline (Butcher, 1962). The present structure is largely the consequence of Variscan north-west-directed transpression. Where simplest, for example in the Storridge road cutting, it is a symmetrical upright anticline, known as the Storridge Anticline (Brooks, 1968). Elsewhere, as in the Malverns, the eastern limb of the fold is either faulted out or overthrust by west-directed thrust sheets. Steepening of the western limb on the west side of the Malverns results in vertical or overturned strata and a broadly monoclinal structure. West-directed thrusting occurs in the Collins Green–Alfrick and Herefordshire Beacon areas. In the latter, this is responsible for the presence of the Precambrian Warren House Formation.

Little can be deduced from the small amounts of Cambrian strata exposed within the district, but their rapid eastward thinning against the Malverns to the south of the district is attributed to Ordovician uplift controlled by north-west-trending faults (Brooks, 1970).

The nature of the western boundary of the Precambrian Malverns Complex, again the subject of much debate in the past, is detailed in Chapter 4. It is locally an unconformity and elsewhere it is a thrust fault. Where it is the latter, the vertical displacement is likely to be small, the greater movement probably being transcurrent. Where the junction is an unconformity, transcurrent movement is probably taken up in the basal Llandovery beds.

The eastern boundary of the Malvern Axis is marked by the East Malvern Fault, a major normal fault throwing down to the east, which forms the western margin of the Worcester Basin and is considerably younger than the reverse faults within the axis itself (p.102).

STRUCTURES OF THE LOWER PALAEOZOIC ROCKS

Although the marine Silurian rocks on and immediately to the west of the Malvern Axis have received much attention (e.g. Butcher, 1962; Phipps and Reeve, 1969), the late Silurian to Devonian Old Red Sandstone facies rocks farther to the west have been almost ignored, and the mapping carried out during the survey has revealed previously unknown details of their structure. West of the Malvern Axis, seismic, magnetic and gravity studies show that the surface of the Proterozoic basement underlying the Lower Palaeozoic is relatively flat, and lies at depths between 1500 m and 2000 m (Figure 33). The nature of the basement remains uncertain, with the probability that the Collington Borehole (Figure 34) did not reach it (p.96). Whether basement rocks akin to the Malverns Complex persist to the west of the Malvern Axis is not clear. It is possible that the rocks exposed are a series of shallow, detached lenses along the Proterozoic suture (p.116) and that Longmyndian basement is present west of the Malvern Axis. This is suggested by the presence of Precambrian Longmyndian rocks (the Huntley Quarry Beds) at outcrop south of May Hill, 20 km south of the district, although the presence of magnetic anomalies indicates that other basement types may exist (p.116).

In the following account, structures are grouped and named according to their predominant trend (e.g. Owen, 1954; Brandon, 1989). North-north-west trends are referred to as Charnoid, north-east trends as Caledonoid, east–west trends as Variscoid and north–south trends as Malvernoid.

The dominant structural trend evident in the Lower Palaeozoic rocks (Figure 32) is Charnoid, along which is a suite of faults and folds. Caledonoid faults and folds occur in the south-west of the district. The main Malver-

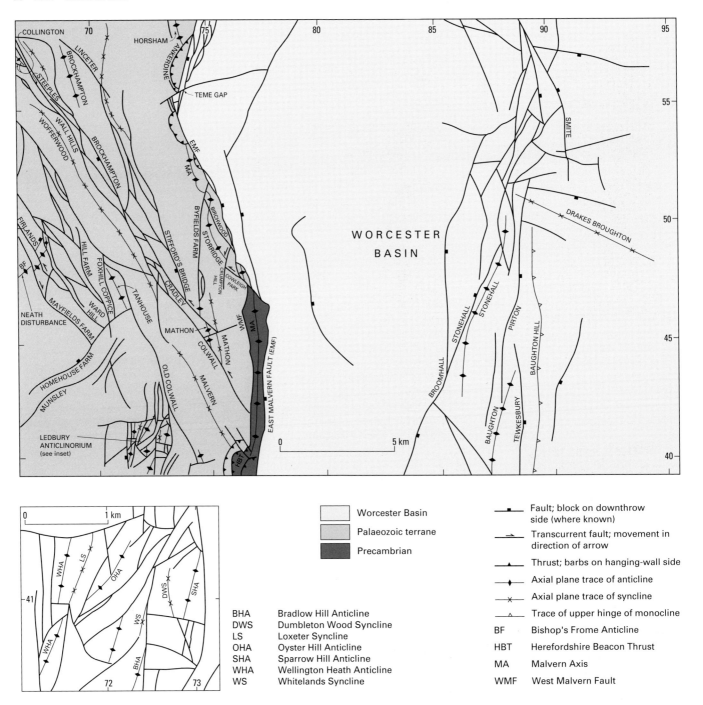

Figure 32 Principal surface structures of the district. Inset: the Ledbury Anticlinorium.

Legend:

☐	Worcester Basin
☐	Palaeozoic terrane
☐	Precambrian

Fault; block on downthrow side (where known)

Transcurrent fault; movement in direction of arrow

Thrust; barbs on hanging-wall side

Axial plane trace of anticline

Axial plane trace of syncline

Trace of upper hinge of monocline

BHA	Bradlow Hill Anticline
DWS	Dumbleton Wood Syncline
LS	Loxeter Syncline
OHA	Oyster Hill Anticline
SHA	Sparrow Hill Anticline
WHA	Wellington Heath Anticline
WS	Whitelands Syncline

BF	Bishop's Frome Anticline
HBT	Herefordshire Beacon Thrust
MA	Malvern Axis
WMF	West Malvern Fault

noid structure is the Malvern Axis, from which the term is derived.

Charnoid structures

A suite of north-north-west-trending faults and folds affects the Silurian and Devonian rocks. The Collington–Brockhampton–Colwall fault system is the most important, although the effects of other faults are difficult to assess because of the lack of stratigraphical markers within the Old Red Sandstone facies.

The Collington Anticline can be traced only for a short distance in the north-west of the district. The Collington Borehole [6460 6100] (Figure 34; Department of Energy, 1978) drilled the structure to the north-west of the district, apparently reaching Precambrian basement at a depth of 1710 m. However, seismic evidence suggests that the basement here lies at a depth of about 1800 m. The basal strata in the borehole may therefore be conglomerates within the Cowleigh Park Formation, of Llandovery age, rather than Precambrian basement (p.14). The Collington Fault cuts the western

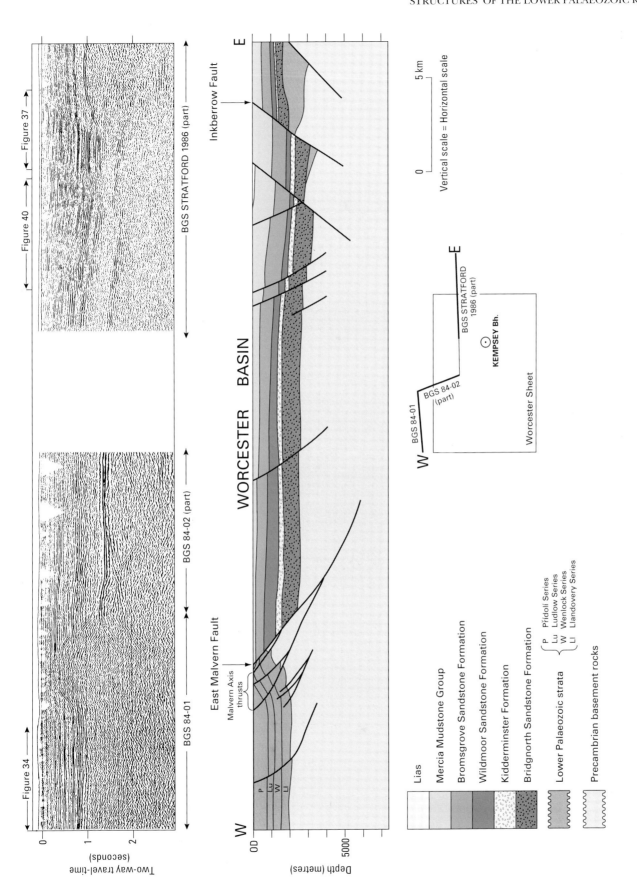

Figure 33 BGS seismic reflection profiles, with simplified geological interpretation, in the vicinity of the northern part of the Worcester district; locations of seismic data used for figures 34, 37 and 40 shown.

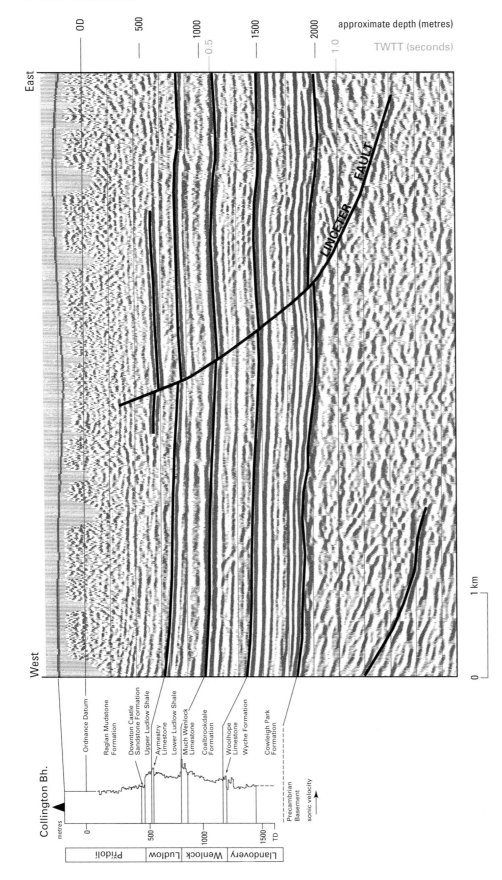

Figure 34   Seismic stratigraphy of the Lower Palaeozoic rocks, related to the sequence in the Collington Borehole. The Linceter Fault shows evidence of syndepositional reverse movement, with later normal displacement. See Figure 33 for location.

limb of the anticline and is a high-angle reverse fault at the surface, flattening to a lower-angle thrust at depth. The fault throws down to the south-west and has a reverse displacement of 100 m; there is also a probable sinistral component of displacement (Brandon, 1989). A parallel fault, the Wall Hills Fault, is also considered to be a sinistral wrench fault, but to have a normal displacement of several hundred metres down to the north-east (Brandon, 1989). These faults appear to converge with the Steeples Fault, and the Collington Anticline is truncated between them. The Steeples Fault is probably the continuation of the Brockhampton–Colwall fault system.

The Brockhampton Anticline lies to the east of the Brockhampton Fault. This is an asymmetric fold in the Raglan Mudstone, with a steep western limb dipping up to 45°, and a gentler eastern limb dipping at 5 to 10°. To the north, in Linceter Brook [6892 5716], dips up to 37° indicate a northerly plunge to the structure; the anticline cannot be traced in the overlying St Maughans Formation to the north. This upwards decrease of structural complexity may be the consequence of differing bulk rock properties. Alternatively, it could indicate a short period of inversion during which the pedogenic Bishop's Frome Limestone was formed. A folding episode such as this may represent a minor precursor to the main Acadian collision event.

The Linceter Fault is mapped to the east of the Brockhampton Anticline and probably corresponds to a fault imaged on seismic line BGS 84-01 (Figure 34). In the subsurface, the fault shows a normal, down-east displacement of the younger Silurian sequence, which switches at depth to a minor reverse down-west displacement in the older part of the sequence. This can be explained as follows. Reverse displacement during deposition of the May Hill Sandstone Group gave a slightly thinner Llandovery sequence on the easterly hanging wall block. Movement on the fault died out thereafter, with younger strata suffering little or no reverse displacement. Subsequent reactivation of the fault in a normal sense, perhaps during Permo-Triassic extension (p.108), produced the observed normal displacements of the younger Silurian strata, but was insufficient to cancel out the larger reverse displacement at depth. The early reverse movement may represent a continuation of the pre-Llandovery tectonism that produced uplift of the Malvern Axis (Brooks, 1970).

East of the Linceter Fault, a broadly synclinal area (Figure 32) passes eastwards into the Horsham Anticline (Phipps and Reeve, 1969, p.16). This is a broad, upright, north-trending fold which can be traced to the River Teme near Hill Top Farm [7375 5811], where it swings south-south-east and is overridden by the Ankerdine Thrust. A fault [7348 5826] on the western limb near Hill Top Farm juxtaposes flat-lying beds to the west against steeply dipping beds to the east.

The Brockhampton Fault was seen in a temporary section in the Ammons Hill railway cutting [7016 5297] (Barclay et al., 1994), faulting St Maughans Formation beds to the west against Raglan Mudstone Formation beds to the east. It is marked by a zone, 40 m wide, of faulted and folded beds. The fault can be traced through a plexus of closely spaced parallel faults in the Cradley area to continue as the Colwall Fault in the Mathon–Colwall area.

The Colwall Fault is a near-vertical sinistral wrench fault. On the basis of the presence of horizontal beds in a borehole [7572 4265] at Colwall Stone, Phipps and Reeve (1969) assumed the axis of the Malvern Syncline to lie on the line of the borehole. On the further assumption that the Malvern and Mathon synclines were formerly co-axial, they calculated a sinistral displacement of about 1200 m on the Colwall Fault at Colwall, decreasing from there south-eastwards towards the Malverns. However, there is a fold closure of the Malvern Syncline, plunging north-westwards and overridden by the Herefordshire Beacon Thrust, east of Oldcastle Farm [757 405]. The synclinal axis trends north-westwards to the west of Mathon [736 448] where the Bishop's Frome Limestone swings round the axis, and thence to Wofferwood in the north-west of the district. It appears, therefore, that the Mathon Syncline does not continue on the south side of the Colwall Fault, nor does the complimentary Mathon Anticline. The amount of transcurrent movement on the fault is therefore not known. Sinistral movement on the Cradley Fault, parallel and to the east of the Colwall Fault, can be demonstrated where the Mathon Anticline is displaced about 300 m near Rowbarrow Wood [748 459]. The anticline is displaced dextrally by an east-north-east-trending fault to the south. The net effect of movement on this and the Cradley Fault has resulted in little demonstrable displacement of the Mathon Syncline (cf. Phipps and Reeve, 1969).

Phipps and Reeve suggested that the Colwall Fault continues to Gardiner's Quarry [765 421] in the Malverns Complex, where there is a zone of intense shearing. There is, however, no evidence of lateral displacement of the western boundary of the complex, as suggested by Phipps and Reeve; they proposed a tectonic junction with the Silurian rocks (the Worcestershire Beacon upthrust) in this area, but there is also no evidence of this. Phillips (1848, p.136) showed the fault (schematically) to become bedding-parallel and die out in the Coalbrookdale Formation. It was exposed in the Colwall railway cutting [7600 4285] (Robertson, 1926) where it faults the Raglan Mudstone about 135 m above its base against the Coalbrookdale Formation, giving a minimum vertical displacement of about 520 m (Phipps and Reeve, 1969).

The Cowleigh Park Fault complex (Phillips, 1848; Groom, 1900; Phipps and Reeve, 1969) appears to be a sinistral wrench system, with Malvern Quartzite of Cambrian age present in the fault zone at Cowleigh Park (p.15), but absent from the Malverns to the south. This implies a minimum sinistral displacement of 500 m. The fault seems to have been at least partly reactivated during Permo-Triassic extension as part of the East Malvern Fault system (p.101).

Other important Charnoid faults include the Stifford's Bridge Fault, proposed to account for the anomalous thinness of the Raglan Mudstone Formation in the Cradley–Suckley area, where only about 300 m are present compared to up to 800 m elsewhere.

Other Charnoid folds include the Tanhouse Anticline, the north-west plunging closure of which is present in the vicinity of Tanhouse Farm [713 470]. This is complementary to the Wofferwood–Malvern Syncline, but is present only in the Tanhouse area, being fault-bounded to the east (by the Old Colwall Fault), to the south (by the Homehouse Farm Fault) and the west (by the Foxhill Coppice Fault). The Foxhill Coppice Fault is the eastern one of a suite of Charnoid faults which bound blocks of differently tilted strata.

Phipps and Reeve (1969) suggested that movement on the Ledbury Fault, a sinistral wrench fault present in the Tewkesbury district (Worssam et al., 1989), was responsible for the folds of the Ledbury Hills. The fault could not be traced north of Ledbury in the vicinity of the junction of the Tewkesbury and Worcester sheets, but the north-south fault mapped as the Ledbury Fault by Phipps and Reeve at Wellington Heath can be traced northwards from there for about 1 km. It may then swing north-north-eastwards, parallel to the strike of the Ludlow rocks, to converge with the Old Colwall Fault, or resume a north-north-west Charnoid trend. In either case, the Raglan Mudstone outcrop has few definite features that can be mapped with conviction which would resolve the problem.

The folds of the Ludlow rocks in the Ledbury Hills, collectively known as the Ledbury Anticlinorium, are a series of faulted anticlines and periclines. They do not have the sigmoidal axial traces suggested by Phipps and Reeve (1969), but the model proposed by these authors, of sinistral transpression between controlling, steep, transcurrent faults, seems plausible. The folds are tight, with parallel limbs dipping generally over 35° and commonly up to 60°. The fold axes are generally faulted, although the Wellington Heath Anticline has a simple fold closure and plunge. The intervening synclines appear to be mostly faulted out, indicating imbricate stacking of the folds by high-angle reverse faulting. The folds are named, from west to east, the Wellington Heath Anticline (Hardie, 1969; the Raven Hill Pericline of Phipps (1957) is probably the northern continuation of this), the Loxeter Syncline, the Oyster Hill Anticline (Hardie, 1969; the Oyster Hill Pericline of Phipps, 1957), the Whitelands Syncline (Phipps, 1957), the Bradlow Hill Anticline (Groom, 1910; the Frith Wood Pericline of Phipps, 1957), the Dumbleton Wood Syncline and the Sparrow Hill Anticline (also the Frith Wood Pericline of Phipps). Section 2 on the 1:50 000 map crosses these structures (close to Section 1 of the Tewkesbury Sheet) and is based on the structural style revealed by the seismic line BGS 84-01, on which all the faults dip eastwards.

Caledonoid structures

A Caledonoid structural grain is present in the south-west of the district, where features in the Raglan Mudstone indicate a north-east–south-west strike, and two similarly trending faults, the Homehouse Farm and Munsley faults are mapped. The former, although apparently truncating the suite of Charnoid faults between and including the Mayfields Farm and Hill Farm faults, is itself truncated, along with the other Caledonoid structures, by the Old Colwall Fault.

The Neath Disturbance is the most southerly of a series of similarly trending major dextral strike slip faults that cross Wales and the Welsh Borderland (e.g. Woodcock, 1984). It is represented by the Bishop's Frome Anticline, which extends into the south-west of the district from the adjoining Hereford district, and can be traced for a short distance east of Bishop's Frome, across the Mayfields Farm and Firlands faults, perhaps with small lateral displacements, but it is truncated by the Ward Hill Fault.

The Malvern Axis is breached by the River Teme at Knightwick, the river taking a right-angled bend to flow through the gap north-eastwards. The East Malvern Fault and the thrusts and outcrops of the Silurian rocks appear to be displaced dextrally, up to 500 m, by a Caledonoid fault in the gap. There is also a change in strike from north-north-west to north-north-east across the gap. Phipps and Reeve (1969) deduced the presence of a major transverse structure here, by the reversal of the plunge of the folds on the Malvern Axis. A Caledonoid fault is mapped in the Mercia Mudstone Group for about 4 km east of the gap. It may continue a short distance into the Raglan Mudstone to the south-west in a small dry valley, but it cannot be traced through the ground to the south-west, where the dominant structural trend is Charnoid. The lineament does, however, approximately align with the Neath Disturbance in the Bishop's Frome area, and may therefore be an isolated remnant of that structure, as postulated by previous authors (e.g. Owen, in discussion of Squirrell and Tucker, 1960; Kellaway and Hancock, 1983; Woodcock, 1984).

Variscoid structures

Brandon (1989) used the term Armoricanoid for similar, east–west structures present in the adjoining Hereford district. Some minor faulting with similar or east-north-east trend occurs in this district. One small reverse fault is exposed at Vinesend Quarry [7517 4760] (Bullard, 1989, p.56), overthrusting to the south. An aeromagnetic and gravity anomaly with this trend in the Swinmore Common–Coddington area [c.690 420] probably marks a basement structure (p.116); a minor east-west fault 1 km south of Coddington may be the surface expression of this structure. Phipps and Reeve (1967) reported facies changes and increased thicknesses of Wenlock and Ludlow strata in an east-west trough (the 'Bradlow Trough') in this area.

Malvernoid structures

The Malvern Axis (p.95) and the East Malvern Fault are the principal Malvernoid structures in the district (Figure 32). Variscan transpression was taken up by westerly directed thrusting along the Malvern Axis (Chadwick and Smith, 1988) to produce the West Malvern Fault and thrusts including the Ankerdine, Ravenshill Wood and Herefordshire Beacon thrusts. Although the term

Malvernoid is used to denote predominantly north-trending structures, the Malvern Axis has a Charnoid trend in the area between Malvern and Knightwick. The East Malvern Fault, the western boundary fault of the Worcester Basin, is a major normal fault which formed by extensional reactivation of the Malvern Axis thrusts during Permo-Triassic times (p.102).

The Malvern Axis has been described previously by Phillips (1848), Groom (1898, 1900) and Phipps and Reeve (1969) and its general nature is summarised on p.102. Its gross structure at depth in the north of the district has been revealed by the BGS seismic lines BGS 84-01/02 (Figure 33) to consist of a series of west-directed, easterly dipping reverse faults, probably linked at depth. The following section gives some details of the structures present at the surface on the axis, described from north to south.

The presence of a small inlier of Malverns Complex rocks thrust over Cambrian Malvern Quartzite at Martley, close to the north of the district (Mitchell et al., 1962) is evidence of thrusting in that area. The Malvern Quartzite is itself thrust over Silurian Raglan Mudstone Formation and Carboniferous Highley Formation strata.

Thrusting in the Collins Green–Ankerdine Hill area is responsible for the outlier of Much Wenlock Limestone at Collins Green (p.31). It overlies mudstones presumed to belong to the Wenlock Coalbrookdale Formation, but if they are of Ludlow age, the contact between them and the overlying limestone is also an east-dipping thrust. The mudstones are thrust westwards over beds of the Raglan Mudstone Formation, the thrust being here named the Ankerdine Thrust. West of Collins Green, clay with sandstone fragments suggests the presence of a thrust-bounded sliver of Llandovery beds between the overlying mudstones and the Raglan Mudstone. The outcrop of the Ankerdine Thrust is not as convolute as indicated by Phipps and Reeve (1969, pl.1); their line marks the toe of solifluction lobes of grey-green Silurian clay.

Ankerdine Hill was described by Phillips (1848, p.149) as an anticlinal feature with a north–south axis bounded by normal faults. Phipps and Reeve (1969, p.16) suggested the presence of two superimposed blocks bounded by low-angle thrusts. Exposures of the Cowleigh Park and Wyche formations on the hill show the beds to be folded into a tight, asymmetric, north-trending, south-plunging anticline which lies above the Ankerdine Thrust. A steeply dipping to overturned sequence from Raglan Mudstone to Aymestry Limestone is present in the footwall of the thrust at Knightsford Bridge [736 557]. Phillips (1848, pp.149, 150) described and illustrated the section. The overturned Aymestry Limestone (p.33) dips 50° towards the east-north-east. Phillips showed an unfaulted, conformable sequence passing up into the Old Red Sandstone, but with the Downton Castle Sandstone absent. Its presence elsewhere throughout the district suggests that it is absent here because of faulting. A similar situation is present immediately south of the Teme Gap, although the structures are apparently displaced dextrally by about 500 m.

East-dipping thrusts are present in the Warren–Ravenshill Wood area [736 537], thrusting beds of the Wyche

and Woolhope Limestone formations on the eastern limb of the Storridge Anticline westwards over the Coalbrookdale Formation (Phipps and Reeve, 1969, pl.3, Section 5). A similar thrust sheet of Coalbrookdale Formation beds is present on the eastern limb of the anticline to the south at Alfrick [745 526].

The Byfields Farm Fault may have acted as a transfer system between the Malvern and Colwall faults. It extends from Old Storridge Common, where it appears to slightly displace the Malvern Axis sinistrally, to the Mathon area, where it lies in the axis of the Mathon Anticline. It is responsible for folding and faulting of the Much Wenlock Limestone in the Whitman's Hill–Storridge area, and for the presence of a small upfold of this limestone at Vinesend [746 473].

The Storridge Anticline in its type area [755 495] is an upright fold with a gentle northerly plunge. The Woolhope Limestone and Coalbrookdale Formation are cut by the Birchwood Fault on the eastern limb of the anticline east of Storridge, bringing in Much Wenlock Limestone. Phipps and Reeve (1969, pl.3, Section 7) invoked low-angle west-directed thrusting here, but the presence of the Birchwood Fault, a steeply dipping normal fault throwing down east, provides a simpler solution. The fault, like several others parallel to the East Malvern Fault, is interpreted as a subsidiary splay of that fault.

The Storridge Anticline is cut and displaced dextrally by the Crumpton Hill Fault, only beds on the western limb of the anticline being present immediately to the east and south of the fault.

Previous interpretations of the structurally complex Cowleigh Park area (Figure 14) were given by Phillips (1848), Groom (1900) and Phipps and Reeve (1969). New data have been provided by boreholes, an excavation of a reservoir [7639 4780] in 1985 and sections exposed during road widening in 1987. A small inlier of Woolhope Limestone occurs in the faulted core of the anticline, with Much Wenlock Limestone and Lower Ludlow Shale on the eastern limb. The Charnoid Cowleigh Park fault system (p.99) follows a linear gravity and magnetic basement anomaly, and, in addition to acting as a Variscan transcurrent fault, it appears to have been reactivated as a subsidiary basin-margin normal fault during Permo-Triassic extension.

From Cowleigh Park south to the district boundary, the Malverns Complex crops out in the Malvern Hills in the core of the Malvern Axis. The dip of the Lower Palaeozoic rocks to the west of the Malverns Complex increases progressively eastwards to vertical and is overturned in proximity to the western boundary.

Details of the western boundary of the Malverns Complex and of the debate formerly generated on its nature are given in Chapter 4. West- or north-west-directed Variscan transpression resulted in a boundary that is locally an unconformity and locally a fault. The fault is here named the West Malvern Fault, in preference to the Worcestershire Beacon Upthrust of Phipps and Reeve (1969), who interpreted the entire length of the junction as a fault. Where an unconformity is present, fault movement was probably taken up by bedding-parallel strike slip in the Llandovery rocks. However, the fact that an unconformity is present locally indicates that there is no great vertical (reversed) displacement on the fault in the West Malvern area.

A synthesis of details listed for localities 10 to 12 (Chapter 4, p.25), together with the record of a fault-bounded block of Cambrian strata in a well at West Malvern (p.15) necessitates the mapping of a complex of closely spaced parallel thrust faults at the western boundary of the Malverns Complex in North Malvern.

The Herefordshire Beacon Thrust (Phipps and Reeve, 1969) is a low-angle fault, thrusting Malverns Complex rocks from the south-east over Silurian strata. A parallel, lower thrust to the west overthrusts Wenlock and Ludlow beds over Ludlow and Přídolí beds which are folded in the axis of the Malvern Syncline, providing evidence of two periods of diastrophism. Minor Charnoid, north-west-trending faults take up some transcurrent movement in this area, apparently displacing the thrusts. A third, higher, thrust to the east of the Herefordshire Beacon Thrust is postulated to account for the presence of the Warren House Formation in the Tinkers Hill–Broad Down area.

The eastern boundary of the Malvern Axis is the East Malvern Fault, which, although of Malvernoid trend, is considerably younger than the reverse faults of the axis, and is described below.

THE WORCESTER BASIN

A limited amount of seismic reflection data provides the principal means of elucidating the subsurface structure of the Worcester Basin. The west–east geoseismic cross-section (Figure 33) illustrates the regional structural configuration of the basin.

Figure 36 Structural contour maps of the district (dashed rectangle) and its surrounds:
a) Base Bridgnorth Sandstone Formation (top pre-Permian basement)
b) Base Sherwood Sandstone Group
c) Base Mercia Mudstone Group
d) Base Lias Group.

Principal structures

The Worcester Basin is a roughly symmetrical graben system (Figure 35), bounded to the west and east by major north–south-trending normal faults. The sedimentary rocks of the basin-fill rest upon Precambrian and Palaeozoic basement. Structural contour maps of the principal stratigraphical horizons within the basin are shown in Figure 36. Apart from a thin cover of Jurassic rocks in the east, and possible (though unproven) late-Carboniferous basal red-beds, the basin-fill within the district is entirely of Permo-Triassic age. Maximum thickness of sedimentary rocks is found in the central and eastern parts of the basin and locally exceeds 3000 m (Figure 36a).

The western margin of the basin abuts the Malverns Axis and is marked by the East Malvern Fault, a major easterly dipping, normal fault. The fault was seen at the following localities, where it has a normal displacement and dips steeply to the east.

1. Temporary section [7430 5785] in which poorly consolidated breccio-conglomerate of the Haffield Breccia to the west is faulted against the Bromsgrove Sandstone.

2. Exposures [7360 5514] in disused railway cutting, Knightwick, in which clays of the Wyche Formation are faulted against the Bromsgrove Sandstone.

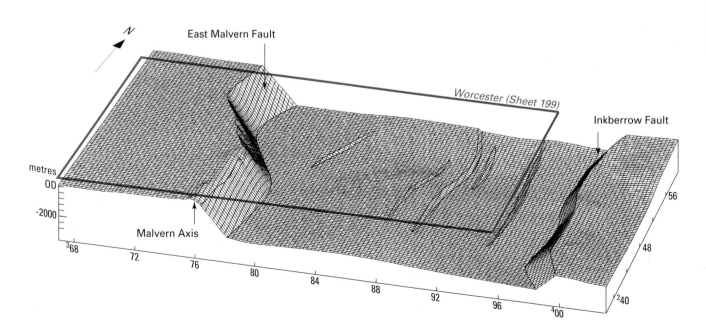

Figure 35 True-scale 3-D view, from the south-south-east, of the top of the pre-Permian basement surface, illustrating the structural architecture of the Worcester Basin.

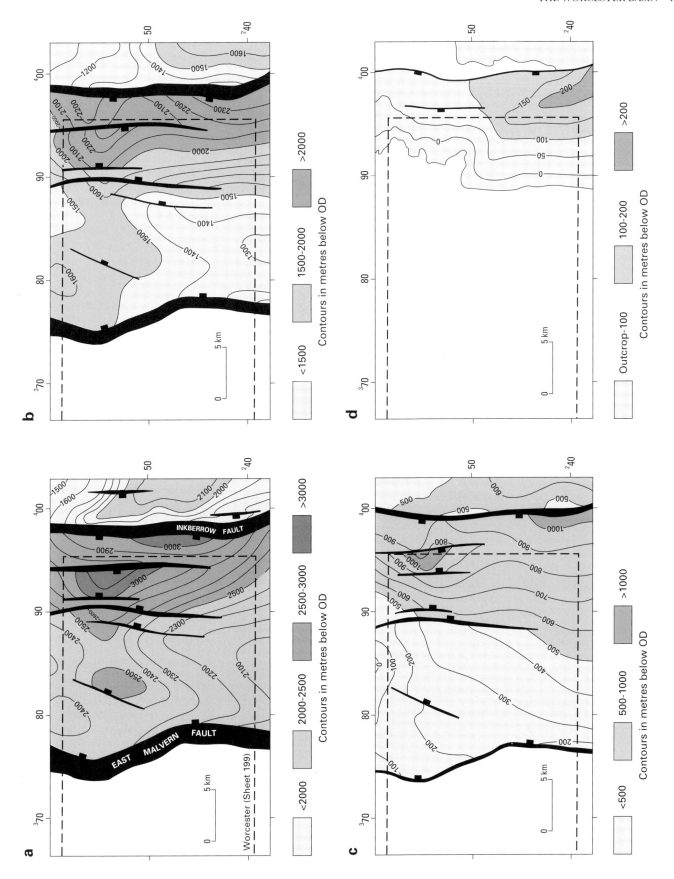

3. Stream bank section [7693 4943] near Storridge in which green mudstones of the Coalbrookdale Formation are faulted against red mudstones of the Mercia Mudstone Group.

4. Exposure [7704 4696] behind car park at North Malvern in which slickensided basal Permo-Triassic pebbly sandstone rests on the surface of the Malverns Complex.

5. Temporary section [7715 4688], exposed in 1982, below the southerly of the two North Malvern quarries, of coarse breccia; boulders of Malverns Complex are banked against the cliffed surface of the Malverns Complex.

The fault was examined in the Colwall railway tunnel [7720 4350] by Robertson (1926), where it consists of a 6 m zone of brecciated rocks of the Malverns Complex in a marly matrix. Antithetic faulting and minor folding are present in the Mercia Mudstone Group beds to the east of the fault, both in the tunnel and in the cutting at the tunnel mouth (Plate 9).

Neotectonic movement of the fault may have been responsible for the Malvern earthquake of 27 September, 1907 (Davison, 1924).

In the subsurface, the East Malvern Fault has a roughly planar fault surface which dips to the east at about $45°$, throwing the Permo-Triassic rocks against Lower Palaeozoic and Precambrian rocks of the Malvern Axis (Figure 33). Taking into account the topography of the Malvern Hills, the throw on the fault at the level of the base of the Permo-Triassic ranges from over 2600 m in the north of the district to over 2300 m in the south (Figure 36a). It is likely (though it cannot be proven, due to the lack of Permo-Triassic strata on and west of the Malvern Axis) that much of this displacement was syndepositional, occurring in Permo-Triassic times.

The eastern margin of the Worcester Basin is marked by several large normal faults which have subplanar surfaces dipping to the west at about $50°$. The most important of these, the Inkberrow Fault (Figure 37), which lies just to the east of the district, has a throw of over 1000 m at the level of the base Permo-Triassic (Figure 36a), which decreases upwards to only 100–200 m at the level of the base of the Jurassic (Figure 36d), indicating over 800 m of syndepositional Permo-Triassic displacement.

Within the basin itself, several smaller normal faults, again with dominantly north–south trends, show evidence of syndepositional movement. There is also a suite of minor faults with an approximate east–west trend. Detailed surface mapping of the many harder, feature-forming beds within the Mercia Mudstone Group has elucidated the shallow structure and revealed a high density of minor faults. The principal structures are confined to a north–south zone of normal faults lying immediately west of the Penarth Group escarpment. The faults form a plexus of north-trending structures of which the Tewkesbury–Pirton-Smite fault system is the most important. Predominantly, they downthrow to the east, and overlie larger normal faults, also downthrowing

east and antithetic to the major eastern bounding faults of the basin. In the north of the district the fault system trends north-north-west. The Smite Fault was the western bounding fault of a subsidiary graben in which the Droitwich Halite was deposited; the Tewkesbury and Pirton faults to the south may also have acted as growth faults and confined halite deposition. The Inkberrow Fault probably formed the eastern bounding fault of the graben, in which is the thickest development of the Mercia Mudstone Group (Figure 36c).

In addition to the complex fault pattern in the Mercia Mudstone Group, there are four significant folds and several minor ones. Three of the structures, the Baughton and Stonehall anticlines, and the Baughton Hill Monocline, have approximate north–south trends parallel to the main rift faults (Figure 32). The Baughton Anticline is well defined by the outcrop of the Arden Sandstone and shows variable dips on the limbs, ranging from 4 to $12°$ and locally up to $20°$. The Stonehall Anticline is a more subdued structure running approximately parallel to, and just east of the Stonehall Fault, and probably formed as a result of drag during movement along the fault. The Baughton Hill Monocline follows the line of the Penarth Group escarpment, and has an easterly dip ranging from 6 to $12°$, compared to a regional dip of about $2°$ to the east. This structure may represent draping over a growth fault at depth.

The Drakes Broughton Syncline trends west-north-west–east-south-east. The maximum dip of the limbs is around $4°$ in the vicinity of the Penarth Group escarpment, decreasing to the east-south-east. This structure is the cause of the sharp indentation of the Penarth Group escarpment south-east of Worcester, and it also has a significant effect on the major north–south-trending faults in this area. The Smite Fault dies out just to the north and its assumed southward continuation, the Pirton Fault, is displaced about 2 km to the west. The general fault pattern in the area shows a curvature roughly parallel to the escarpment. In the Upper Wolvercote area [914 508], the northern limb of the syncline and a complementary minor anticline are cut by a number of small faults of variable trend.

The basin-fill

The Permo-Triassic basin-fill comprises, in ascending order, the Bridgnorth Sandstone Formation, the Sherwood Sandstone Group (consisting of the Wildmoor Sandstone Formation, the Kidderminster Formation and the Bromsgrove Sandstone Formation), the Mercia Mudstone Group and the Penarth Group. Each of these units has its own characteristic seismo-stratigraphic motif (Figure 38). Seismic motifs are not uniquely diagnostic of lithology, since they are to some extent the product of the seismic source (compare the dynamite and Vibroseis data in Figure 33) and limitations of the technique itself (for example a ubiquitous downwards decrease in fine resolution). However they are of use in deducing the general properties of a seismic sequence, such as the presence or absence of significant lithological heterogeneity and the gross style of sediment deposition.

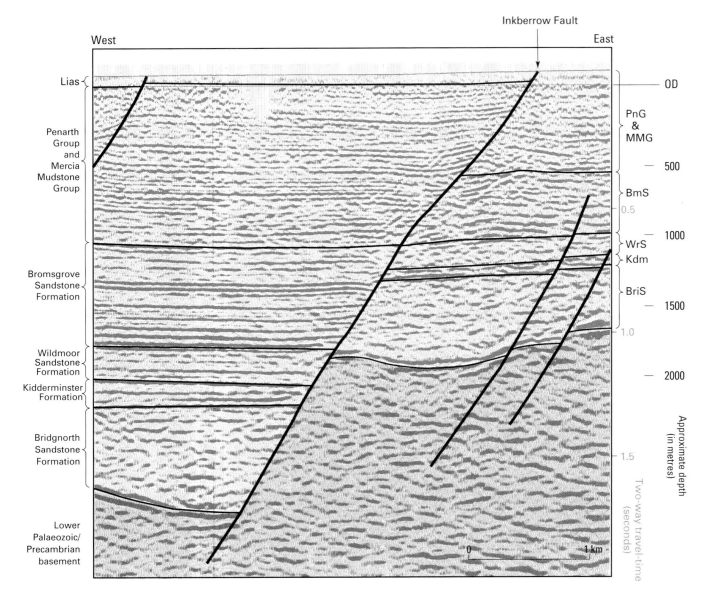

Figure 37 Migrated seismic profile across the Inkberrow Fault. Syndepositional movement is indicated by markedly thickened strata on the downthrown side of the fault. See Figure 33 for location.

The Bridgnorth Sandstone Formation has few prominent or coherent seismic events, perhaps indicative of a generally uniform sandstone sequence with little lithological variation.

The Kidderminster Formation has a much more seismically reflective character, which varies across the basin. In the west (Figure 38a), close to the East Malvern Fault, reflections are of high amplitude, and structurally complex, with a markedly hummocky texture. This may be indicative of large complex (possibly lenticular) sedimentary bodies, perhaps comprising conglomeratic alluvial fans derived from the contemporary fault scarp of the East Malvern Fault. Farther east (Figure 38b) the hummocky reflections die out, to be replaced by more uniform horizontal events, suggesting a more distal depositional system with simpler, flatter sedimentary units.

The Wildmoor Sandstone exhibits a similar seismic character to the underlying Bridgnorth Sandstone, few coherent reflections suggesting a rather uniform sandy lithology. The unit thins markedly eastwards (Figure 33) with a slightly diachronous top, younger in the west, indicating that sandstone deposition continued longest in the west, close to the East Malvern Fault. A slight eastward increase in reflectivity also suggests a more distal depositional environment, the incoming of more argillaceous beds, giving rise to greater acoustic heterogeneity.

The Bromsgrove Sandstone Formation is markedly more reflective than the underlying sandstone units, with numerous prominent flat reflections. This is consistent with the presence of interbedded arenaceous and argillaceous layers proved in boreholes. The seismic signature of the formation is laterally variable, with the

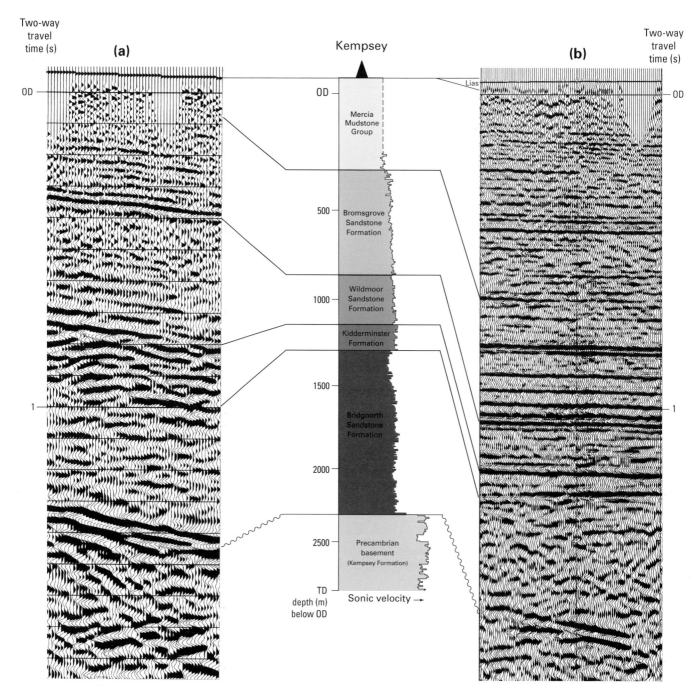

Figure 38   Seismic stratigraphy of the Worcester Basin related to the Kempsey Borehole.
a) Near western margin of the basin (dynamite seismic line BGS 84-02).
b) East-central part of basin (Vibroseis seismic line BGS Stratford 1986).

sharpest variations occurring across faults, suggesting that contemporaneous faulting influenced facies distribution. Within fault-blocks, individual reflection packages are commonly of considerable extent, indicating stratigraphy which is laterally rather uniform.

The Mercia Mudstone Group constitutes a highly seismically reflective sequence, with abundant closely spaced coherent events of considerable lateral continuity. This type of seismic signature is consistent with a thick se-

quence of thinly interbedded silty and muddy strata. Higher-amplitude events probably correspond to evaporitic layers with local developments of halite. As with the Bromsgrove Sandstone Formation, there is a lateral consistency of seismic stratigraphy within individual fault blocks, stratigraphical variation predominantly occurring across faults.

Isopach maps of the concealed Permo-Triassic formations are illustrated in Figure 39 (the Mercia Mudstone

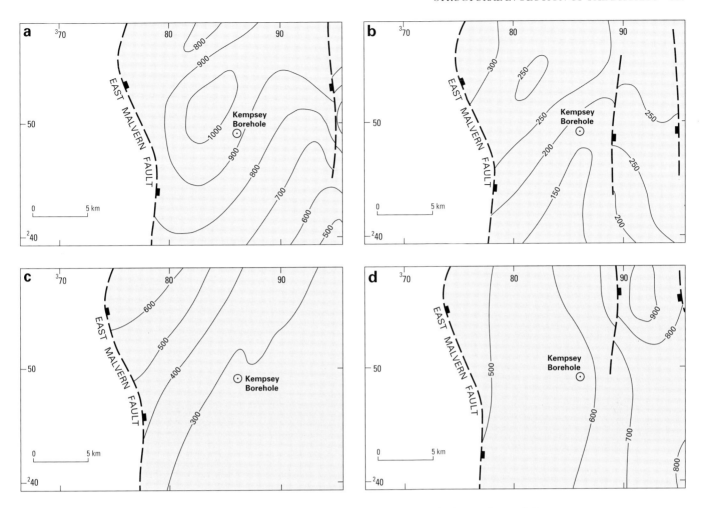

Figure 39 Isopachs of preserved thickness of the concealed Permo-Triassic formations of the district. Dashed lines denote syndepositional normal faults.
a) Bridgnorth Sandstone Formation
b) Kidderminster Formation
c) Wildmoor Sandstone Formation
d) Bromsgrove Sandstone Formation.

Group and Penarth Group isopachs are not shown, since they crop out over much of the basin and have suffered considerable erosion). The maps show preserved compacted thickness, rather than the original depositional thicknesses. In spite of this, salient depositional features can be seen. All the formations show evidence of syndepositional faulting. The lowest three formations (Figure 39a, b, c) show significant west or north-westward thickening, indicating that normal displacement on the East Malvern Fault, with westward tilting of its hanging wall block, was a primary influence on basin subsidence. In contrast, the Bromsgrove Sandstone Formation (Figure 39d) has a depocentre markedly farther east, towards the centre or eastern part of the basin. This may indicate that the Inkberrow Fault became the dominant depositional influence. It is more likely however, that fault-controlled basin subsidence was becoming progressively less important, as it was replaced by a more regional, basin-centred subsidence. Deposition of the Mercia Mudstone

Group conformed to this general trend, with continued active faulting, but within a regional basin-centred subsidence pattern (Figure 33).

A minor angular unconformity within the upper part of the Eldersfield Mudstone Formation is visible on the seismic data (Figure 40). The unconformity is not recognised at outcrop and its origin is uncertain. Deposition at that time was either subaerial or in very shallow water and it is possible that localised uplift and erosion were triggered by either slight changes in regional tectonic stress or perhaps by minor contemporary salt movements in the underlying strata.

STRUCTURAL EVOLUTION OF THE DISTRICT

The principal structural trends of the district have their origins in the Proterozoic during the accretion and assembly of terranes on the northern margin of the East-

West

East

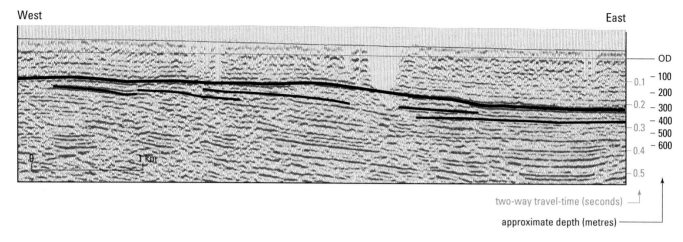

two-way travel-time (seconds) ↲

approximate depth (metres)

Figure 40 Part of seismic line BGS Stratford 1986, showing minor angular unconformity within the upper part of the Eldersfield Mudstone Formation. Lower reflectors pinch out beneath the prominent upper reflector.

ern Avalonian continent (e.g. Soper and Woodcock, 1990). Lower Palaeozoic tectonic history is essentially one of repeated reactivation of the Proterozoic basement structural template, with movement on master faults producing faulting and folding in the cover rocks as the Avalonian and Laurentian continents collided following closure of the intervening Iapetus Ocean.

The Malvern Axis may have been an active structural feature in Cambrian times; an unconformity separates the Comley Series Hollybush Sandstone from the Merioneth Series White-leaved Oak Shale, with no St Davids Series beds present in the Malvern area. Ordovician tectonism resulted in the uplift and removal of all but remnants of the Cambrian and early Ordovician (Tremadoc Series Bronsil Shale) strata within the district. Sedimentation resumed in mid-Llandovery times, when the Malvern Axis appears to have formed the western margin of the Midland Platform and the eastern margin of the Welsh Basin. The presence of coarse conglomerates of reworked Malverns Complex in the basal beds of the Cowleigh Park and Wyche formations points to a shoreline along the west side of the Malverns. Rapid eastward thinning of constituent formations of the Silurian in the vicinity of the axis (Phipps and Reeve, 1967), for example by contemporaneous reverse movement on the Linceter Fault (Figure 34), points to its continuing effects.

The closure of Iapetus and the collision and docking of the Avalonian and Laurentian terranes was marked by dextral transpression on (possibly new) Caledonoid structures and probable sinistral transpression and uplift of the Malvern Axis. Collision marked the onset of continental red-bed Old Red Sandstone sedimentation in the Přídolí, the molasse deposits of the Acadian Orogeny. Coarse-grained lithic sandstones present locally at the base of the Raglan Mudstone suggest renewed uplift and sediment dispersal from the Malvern Axis. Regional uplift in Middle Devonian times resulted in the removal of Pragian–Emsian strata from the district and the absence of any Middle Devonian strata from the Welsh Borderland.

From late Carboniferous times onwards there is sufficient evidence to piece together a story of crustal compression and extension with repeated reactivation of both the Malvern Axis and a similar, concealed Malvernoid feature to the east of the district. This structure, which is believed to subcrop about 2 km west of Alcester, is here termed the Alcester Thrust (Figure 41).

It is likely that Carboniferous strata were deposited over much of the district (Figure 41a), but in late Carboniferous times, Variscan transpressive tectonic stresses had a progressively increasing structural influence. The precise nature and direction of the Variscan stress fields is uncertain, but there was undoubtedly a significant component of east–west crustal shortening. Folding and westerly directed reverse faulting and thrusting occurred along the Malvern Axis, with easterly directed thrusting along along the Alcester Thrust beneath the eastern margin of the Worcester Basin (Chadwick and Smith, 1988). These movements caused Precambrian basement rocks to be thrust against and over Lower Palaeozoic strata, with major uplift of a fault-bounded interior tract (Figure 41b), having an overall geometry akin to a large positive 'flower-structure' (Harding, 1985). Much of the Carboniferous strata deposited in the district were removed at this time, erosion exposing large areas of Precambrian basement rocks within the uplifted interior tract (Figure 41c).

In Permo-Triassic times, tensional tectonic stress fields led to significant east–west crustal extension, with reactivation of the Variscan reverse faults as major normal faults (Figure 41d). Thus the reverse faults of the Malvern Axis were reactivated in a normal sense to produce the East Malvern Fault. Similarly, beneath the eastern margin of the basin, the Alcester Thrust was reactivated in a normal sense to produce the Inkberrow and adjacent faults. This marginal normal faulting caused the previously uplifted interior tract to collapse and the Worcester Basin to form, basin-fill strata resting unconformably upon Precambrian rocks. This preferential subsidence of a previously elevated region is a good example of negative structural inversion (Chadwick and Smith, 1988).

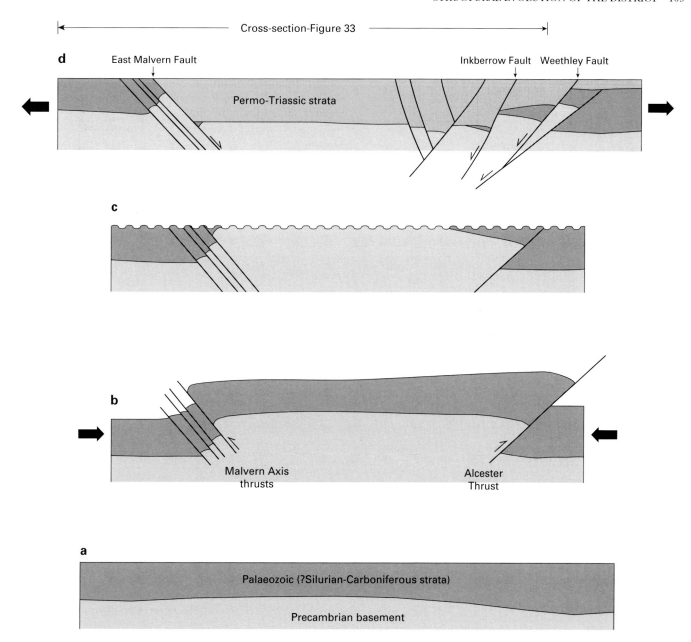

Figure 41   Restored sketch cross-sections to illustrate the post-Carboniferous structural evolution of the district.

a) Late Carboniferous times, prior to Variscan movements

b) Variscan compression leads to crustal shortening and reverse faulting along the Malvern Axis and thrust marked 'F'

c) End-Carboniferous to early Permian times — regional erosion, particularly of uplifted central tract

d) Permo-Triassic tension leads to extensional reactivation of Variscan fractures, with basin subsidence controlled by major basin-margin normal faults.

Crustal extension occurred intermittently during Permo-Triassic times; episodes of active extension corresponded to periods of syndepositional normal faulting, and the rejuvenation of fault-scarp topography which may have enhanced deposition of more arenaceous sedimentary sequences (e.g. the Kidderminster Formation).

In later Triassic times a period of relative tectonic quiescence is likely to have been associated with reduced topographic relief and deposition of the progressively more argillaceous Mercia Mudstone Group.

The structural evolution of the district since Permo-Triassic times is poorly understood because younger rocks have been largely removed by erosion. It is likely, however, that a considerable thickness of Jurassic and Cretaceous strata was deposited, particularly over the Worcester Basin. Studies of apatite fission-track data from northern England (Lewis et al., 1992) indicate that maximum post-Variscan depth of burial was attained in earliest Cenozoic times. It is likely that this was also the case in the Worcester district. Sonic and density log values from the Kempsey Borehole indicate an eroded overburden of some 1400 m. This suggests a basinwide figure of about 1000 m of eroded post-Triassic strata, with more in the basin depocentres. West of the Malvern Axis the degradation of heavy minerals in the Lower Old Red Sandstone indicates a burial depth of less than

1000 m (Morton, 1988). Sonic log values from the Collington Borehole, however, indicate rather higher figures for the thickness of former post-Lower Devonian strata. Here, however, it is uncertain whether maximum burial depths were attained in early Cenozoic times, or much earlier, perhaps in the late Carboniferous, immediately prior to Variscan uplift.

In Cenozoic times the Worcester Basin suffered major uplift and erosion. Recent work in northern England (Chadwick et al., 1994) suggests that Cenozoic uplift comprised two distinct components; a regional uplift perhaps related to emplacement of the Scottish Tertiary Igneous Province, with superimposed, uplifts due to more localised basin inversion. Westward shallowing of

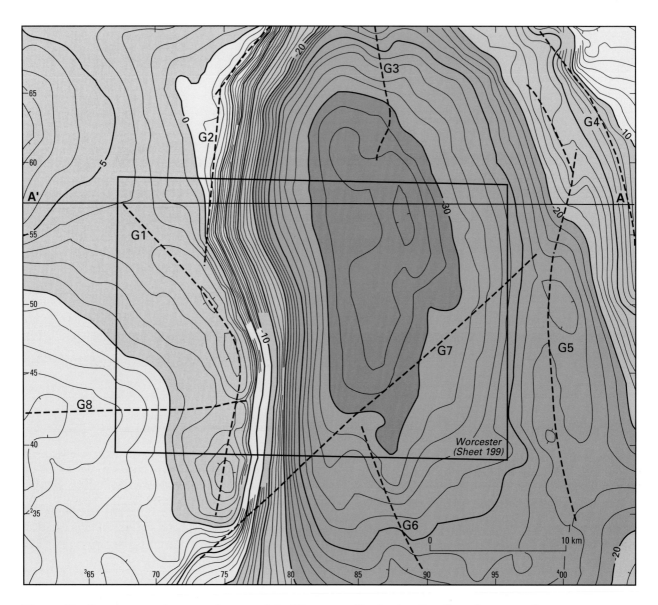

Figure 42   Bouguer gravity anomaly map of the Worcester district (outlined) and surrounding area with contours at 1 mGal intervals. Density for data reduction 2.50 Mg/cm³. Anomalies G1–G8 referred to in text.

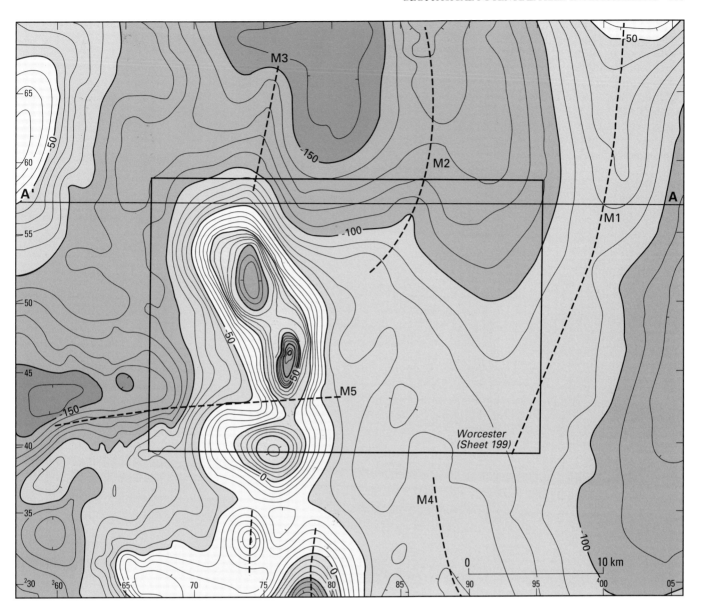

Figure 43   Aeromagnetic anomaly map of the district with contours at 10 nT intervals. Anomalies M1 to M5 referred to in text.

the base of the Mercia Mudstone Group (Figure 33) indicates that uplift was greatest close to the basin's western margin, consistent with basin inversion controlled by reversed movement, down to the west, of the East Malvern Fault. Such inversion probably resulted from transpressive stresses arising from Alpine collisions to the south. Subsequent to basin inversion, the district as a whole has continued to suffer regional uplift and eastward tilting to the present day.

GEOPHYSICAL POTENTIAL FIELD INVESTIGATIONS

The regional gravity and aeromagnetic data provide additional information on the concealed geological struc-

tures of the district. The following account summarises the results of the geophysical investigations described in Cornwell (1992) and discusses the geological significance of the main geophysical anomalies. Bouguer gravity anomaly and aeromagnetic data are presented in Figures 42 and 43 respectively; Figure 44 is a shaded colour plot of the gravity data for a wider area, placing the district in its regional context.

Gravity and aeromagnetic profiles across the Worcester district and parts of the adjacent districts are shown in Figure 45, together with an interpreted model for the upper crust. The densities used in the model and listed in the caption are based on results reported by Cook and Thirlaway (1955) and Brooks (1968) and from borehole logs. These indicate not only density contrasts between the Permo-Triassic and Palaeozoic rocks, but sig-

Figure 44 Shaded relief plot of gravity data illuminated from the west. Colour changes occur at intervals of 12.5 mGal with the red (positive values)/ blue change at 0 m Gal. Rectangle indicated Worcester mapsheet.

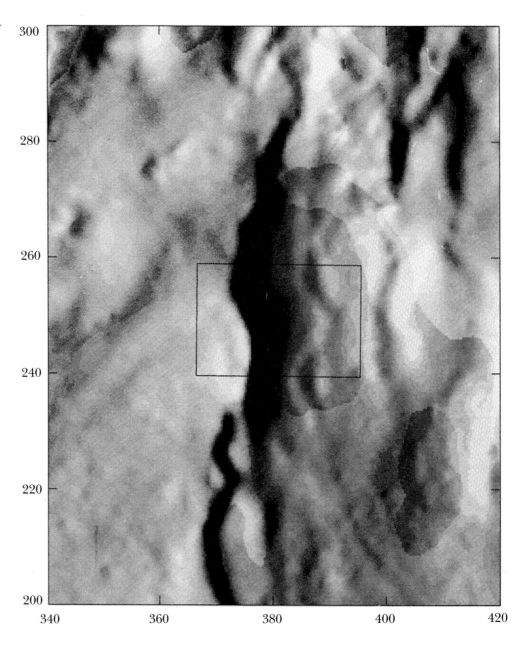

nificant contrasts between the Lias, Mercia Mudstone and Sherwood Sandstone groups. The main sources of magnetic anomalies are the rocks of the Precambrian Malverns Complex. The results of a recent examination of the magnetic susceptibility of these are discussed below.

Worcester Basin

The gravity low associated with the Worcester Basin is a particularly pronounced feature of the gravity map of central England due to its large amplitude and sharp margins (Figure 44; Lee et al., 1990). The feature was interpreted by Cook and Thirlaway (1955) and Brooks (1968) to provide the first estimates of the thickness of sedimentary rocks in the basin, Brooks' figure of about 2.3 km (7500 feet) being largely confirmed by the Kempsey Borehole and seismic reflection surveys (Whit-

taker, 1980, 1985; Chadwick, 1985, Chadwick and Smith, 1988). The gravity map shows the low to be strongly asymmetrical for much of its extent, with the steepest gradient to the west across the basin-margin East Malvern Fault and minimum values about 10 km to the east. The lower gradients to the east are due to the absence of a single eastern boundary fault, to the incoming of intermediate density (2.50–2.55 Mg/m^3) Carboniferous and Devonian rocks below the Mesozoic, and to the eastward thinning of the lower density Permian and Triassic sandstone sequence. The model for the Worcester Basin in Figure 45 is based on evidence of seismic reflection surveys (Figure 33) and reproduces the observed Bouguer gravity profile with little modification. However, a higher-density basement is required beneath the eastern part of the basin and this is interpreted, in part, as being the magnetic body responsible for the anomalies to the east.

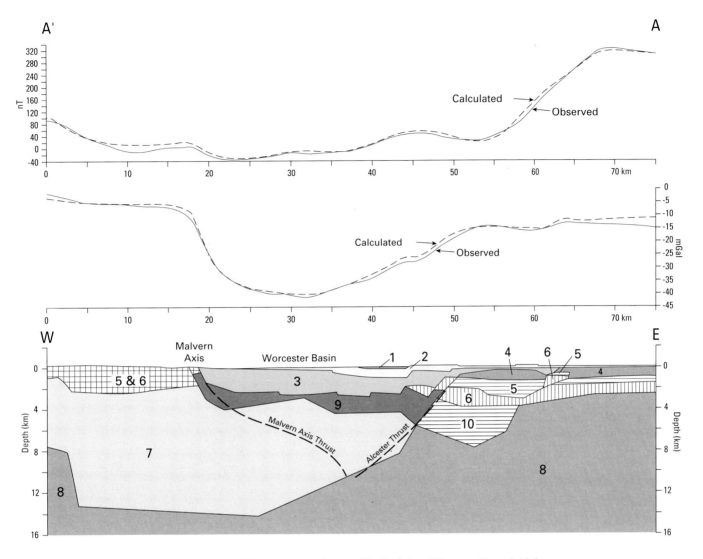

Figure 45 Observed aeromagnetic and Bouguer gravity profile AA′ (see Figures 42 and 43 for location) and calculated profiles for the model shown. Key to model:

1 — Lias (density 2.56 Mg/m^3, susceptibility 0);

2 — Mercia Mudstone (2.46, 0);

3 — Permian sandstones and Sherwood Sandstone (2.27, 0);

4 — Coal Measures (2.50, 0);

5 — Lower Palaeozoic (upper part) (2.55, 0);

6 — Lower Palaeozoic (lower part) (2.65, 0 lower part 0.03);

7 — weakly magnetic basement including Malverns Complex (2.70, 0.012);

8 — magnetic basement (2.72, 0.03);

9 — non-magnetic basement (2.70, 0);

10 — dense basement (2.80, 0).

Background fields: aeromagnetic -110 nT. gravity 8 mGal.

Malvern Axis Thrust and Alcester Thrust are thrusts indicated by Chadwick and Smith (1988).

Superimposed on the broad low over the basin are low-amplitude gravity anomalies (such as G3, G4, G5, G6 in Figure 42) that appear to be related to Mesozoic structures and are probably partly due to density contrasts between the Lias, Mercia Mudstone and Sherwood Sandstone groups. Examples of these in the eastern part of the Worcester district and the adjacent area are illustrated in a plot of residual anomalies (Figure 46) based on gravity data reduced using a density of 2.40 Mg/m^3. A group of anomalies in the east is due, partly at least, to the presence of higher density Lias Group rocks and the north–south elongation corresponds to the trend of

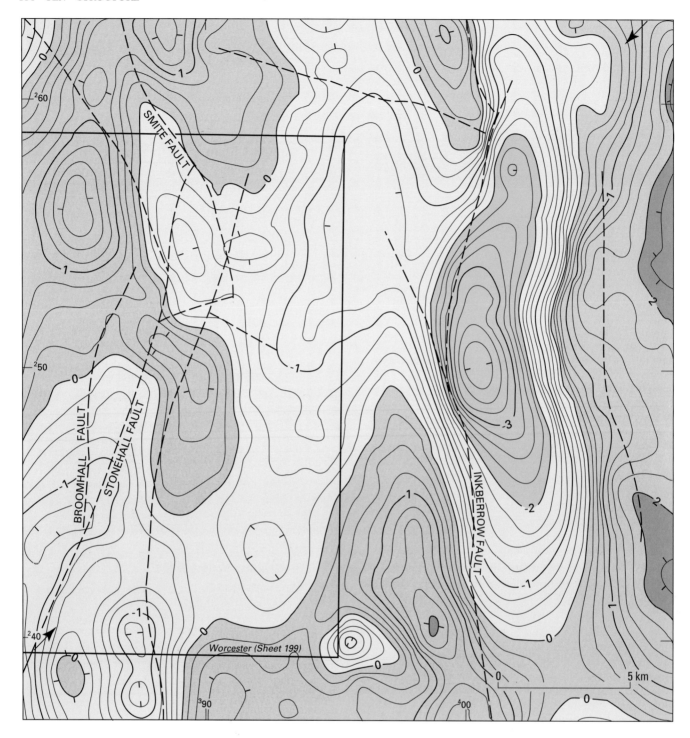

Figure 46   Residual Bouguer anomaly for part of Worcester Basin with contours at 0.25 mGal intervals (0 mGal indicated). Surface faults are indicated by broken lines and the course of possible NE–SW lineament by arrows.

the Inkberrow Fault. A north-north-west-trending lineament aligns with, and may be related to, the synclinal axis in the Inferior Oolite Group in the south-east of the district (p.83). The gravity data also give some subtle indication of a north-east–south-west zone (G7 in Figure 42) apparently displacing north–south-trending anomalies. The projection of this zone coincides with the southern limit of exposed Malverns Complex rocks, but corresponds with only minor faults affecting the Permo-Triassic.

BASEMENT STRUCTURE

The aeromagnetic map (Figure 43) reveals the presence of several low amplitude anomalies which must originate in the pre-Mesozoic basement, and suggest that this largely comprises magnetic rocks. These may be part of the Precambrian Kempsey Formation, proved in the Kempsey Borehole, although the core samples are only weakly magnetic. In the Netherton Borehole (Sheet 200), the source of the magnetic anomalies must lie beneath the proved Lower Palaeozoic rocks.

In the interpreted profile (Figure 45), magnetic rocks are indicated both beneath the Worcester Basin and the Palaeozoic rocks in the west. Farther south, however, the evidence becomes stronger for the existence of magnetic basement beneath the basin and for its disappearance west of the Malvern Axis. The evidence from the interpreted profile suggests that it is probable that basement rocks to the east of the Worcester Basin differ from the exposed Malverns Complex in being more magnetic and having a slightly higher density.

The magnetic map (Figure 43) shows three elongated anomalies indicating possible structures in the basement beneath the Worcester Basin. The most pronounced of these is associated with the Inkberrow–Haselor House Axis (M1 in Figure 43) and is probably due to rocks displaced by the Weethley Fault. North of the Worcester district, this magnetic feature extends to the Lickey Hills area, where its coincidence with the outcrop of the Ordovician Barnt Green Volcanic Group (Old et al., 1991) suggests that these rocks may be related to the source of the anomalies. The southern part of the magnetic anomaly diverges from the Inkberrow–Haselor House structure to trend south-south-west across the eastern part of the district, independently of any known structure.

A second linear magnetic feature (M2 in Figure 43) is not associated with any known structure, although northwards it develops into a large magnetic anomaly along the major fault forming the western boundary of the South Staffordshire Coalfield. The weak feature indicated by M4 becomes more pronounced southwards, but its origin is also unknown.

The Malvern Axis

The Malvern Axis is associated with pronounced gravity and magnetic anomalies. The average density of the rocks of the Malverns Complex (2.70 Mg/m³) is not unusually high for rocks of Precambrian age and the forms of the gravity anomalies result from the effect of low-density rocks of the Worcester Basin to the east and the intermediate density Lower Palaeozoic rocks to the west. The magnetic anomalies are more symmetrical, with values decreasing to the east and west of the narrow high over the axis. In an attempt to identify the sources of the anomalies, magnetic susceptibility measurements were made of rocks in the Malverns Complex, Warren House Formation and adjacent Palaeozoic sequence. The more magnetic rocks appear to be the amphibolites, particularly the diorites, of the Malverns Complex and the

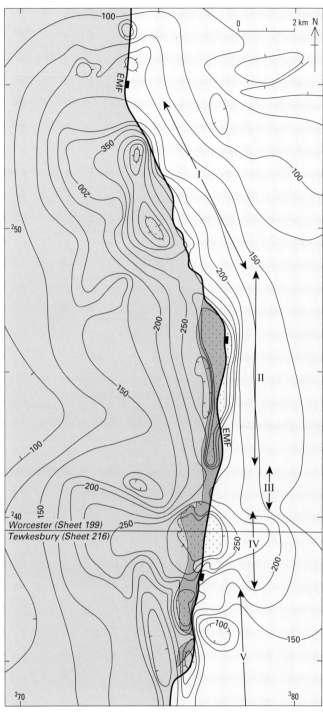

Permo-Triassic strata

Palaeozoic rocks

Precambrian basement (Malverns Complex)

Areas of intense anomalies

East Malvern Fault (EMF)

Figure 47 Magnetic anomaly map (vertical component) based on the ground survey results of Brooks (1968) with contours at 25 nT intervals except for areas of intense values (stippled). Outcrop of Malvernian rocks shaded. Zone I to V described in text.

rocks of the Warren House Formation, but their average value is insufficient to explain the observed magnetic anomalies. The average magnetisation, equivalent to a susceptibility of 0.012, used in Figure 45 is that adopted by Brooks (1968). The most consistently high susceptibility values (about 0.01, with some very high values of more than 0.1) were observed for microdioritic intrusions in the Malverns Complex. If these Precambrian intrusions are responsible for the anomalies, the implication is that not only must they be very numerous, but their magnetic signature might be difficult to distinguish from that of the microdiorite dykes, which are also magnetic, intruding Cambrian rocks of the south Malverns.

Brooks (1968) demonstrated the association of magnetic anomalies with the Malverns Complex. However, where the Malverns Complex is exposed the anomalies were found to be too variable to indicate on the contour map. North of Grid line 355N the line of the concealed Malvern Axis is indicated by the pronounced gravity gradient and a weak magnetic anomaly (M3 in Figure 43), the latter indicating the subsurface continuation of magnetic Malverns Complex rocks, although probably of more limited extent. South of this line the main evidence for variations in the nature of the Malvern Axis is apparent from the magnetic map and five zones are recognised (I to V in Figure 47). The magnetic anomalies forming zone I provide clear evidence of the concealed extension of the Malverns Complex in the Storridge Anticline which Brooks (1968) estimated to lie at a depth of about 300 m in the core of the structure. The magnetic data confirm the presence of the north-west-trending Cowleigh Park sinistral fault system which truncates the Malverns Complex at the surface at North Malvern.

In the northern part of the exposed Malverns Complex in the Malvern Hills (Zone II) there is evidence, in the form of a high-amplitude aeromagnetic anomaly superimposed on a longer wavelength anomaly, that the surface rocks overlie a deeper-seated core. Brooks' ground survey (1968) revealed a complex series of anomalies over the exposed rocks and, locally, negative anomalies immediately to the west. This second feature, which is not apparent in the aeromagnetic data because of loss of resolution, provides evidence of the locally overthrust western boundary of the Malverns Complex. The aeromagnetic survey data indicate that the western boundary of the magnetic rocks dips less steeply at intermediate depths (2 to 6 km) and more steeply at greater depth. There is also evidence that the outcrop of the Malverns Complex occurs slightly to the east of the main magnetic ridge at depth.

The ground magnetic anomalies decrease in width southwards to about 340N, near Little Malvern (Zone III), although the outcrop width remains unchanged. This may be due to a decrease in the proportions of magnetic rocks; the granite and diorite exposed in the road cutting [769 437] at Wyche, for example, produce consistently low susceptibility values. South of 340N the magnetic anomaly pattern changes abruptly with the appearance of a large-amplitude feature from near-surface sources and a more pronounced longer-wavelength component (Zone IV). The former coincides with the outcrop of the Warren House Formation and suggests that these rocks may be responsible for at least part of the deeper magnetic zone. In fact, the rocks are only moderately magnetic at outcrop, but this may be due to weathering, and the ground magnetic anomaly coincident with the outcrop indicates high values en masse. The truncation of the magnetic anomaly along the east–west 340N grid line, along with the gravity feature G8 (Figure 42), probably indicates the presence of a Variscoid basement fault. The southern section of exposed Malverns Complex rocks (Zone V, Sheet 216) is characterised by small, elongate anomalies in the ground magnetic data, which, together with a negative anomaly to the west and the absence of a gravity high and of a longer wavelength magnetic anomaly, suggests that (magnetic) Malverns Complex rocks occur only as a series of near-surface thrust slices in this area. Both ground and airborne magnetic data indicate a second anomaly 2km to the west, which continues as a broader anomaly in Zone IV to the north. This may be caused by dykes in the Cambrian rocks, although Brooks (1968) discounted these intrusions as a source of major anomalies.

To the south of the district the line of magnetic anomalies associated with the Malvern Axis is continued by the well-defined anomaly at Newent (cf. Brooks, 1968). The anomaly is larger in area and amplitude than that associated with the exposed Malverns Complex rocks, but its source is concealed by the Permo-Triassic sedimentary cover. This magnetic evidence, however, suggests that the Malvern Axis in this area lies just to the west of the Hartlebury Fault of the Tewkesbury district. The main Bouguer gravity anomaly in this area (Figure 44) is associated with the margin of the Mesozoic rocks.

Palaeozoic rocks

There are few discrete geophysical anomalies associated with the outcrop of the Palaeozoic rocks immediately to the west of the Malvern Axis. The relatively low-density Cambrian, Silurian and Devonian rocks are associated with gravity lows and have low magnetic susceptibility. Magnetic anomalies appear to be due to Precambrian basement rocks in addition to the postulated Longmyndian rocks (p.95), which elsewhere are known to be only weakly magnetic. Gravity values decrease towards the south-west of the district to an approximately east–west-trending minimum at about grid line 242N (G8 on Figure 42). This, together with a similarly trending magnetic low (M5 on Figure 43), indicates a basement structure which may have exerted control on Palaeozoic deposition, lending support to Phipps and Reeve's (1967) suggestion of a thickened Wenlock and Ludlow sequence ('the Bradlow Trough') in this area (p.100).

ELEVEN

Economic geology

WATER RESOURCES

The district lies predominantly within the River Severn catchment and River Avon sub-catchment, drainage being generally towards the south (Figure 2). The north-west of the district is drained by the River Teme, which joins the Severn near Worcester. The western margins of the district are drained in a southerly direction by the Frome and Leadon, tributaries of the rivers Wye and Severn respectively, and Cradley Brook which flows northwards, becoming Leigh Brook before it joins the Teme at Leigh. The Avon subcatchment includes the Bow Brook, which joins the Avon near Eckington.

Average rainfall ranges from about 650 mm in areas of low elevation along the Severn valley to over 700 mm to the west of Great Malvern and 750 mm in the north-west of the district. Average annual potential evapotranspiration is about 510 mm over much of the district but is likely to be slightly less over the higher ground in the west.

Worcester was once dependent on the River Severn, supplemented by tributary streams and wells for its water; piped supplies of Severn water have been available since the 17th century (Richardson, 1930), and public water supply continues to be drawn from the Severn near Worcester.

Public supplies to rural communities were, in the past, generally obtained from springs, wells and boreholes but many were unreliable, contaminated or of poor quality (Richardson, 1930, 1935). Virtually all of these small sources in the area to the south of Worcester fell into disuse during the 1960s and early 1970s. The Malvern area was originally supplied by springs and capture trenches in the Malvern Hills, but the available quantity was frequently inadequate. Although these supplies are still used at times of heavy demand, the lower-lying parts are now normally supplied with Severn water. Higher areas receive water from groundwater sources near Bromsberrow to the south of the district. The rural areas north of Worcester are supplied from groundwater sources in the Bromsgrove–Redditch area, north of the district. Current groundwater abstraction licences are all for agricultural purposes except two for industrial cooling and two for bottling (Table 16). There are no public supply groundwater sources within the district.

The Sherwood Sandstone Group offers the most favourable potential for groundwater development. However, it is largely concealed at depth beneath the Mercia Mudstone Group and has been little utilised. Although there are no other major aquifers within the district several formations provide valuable local supplies.

Table 16 Groundwater abstraction licence data for the Worcester district.

Aquifer	Agriculture				Industrial				Totals	
	Excluding spray Irrigation		Including spray Irrigation		Bottling		Cooling			
	m³/year	No. licences	m³/year	No. licences	m³/year	No. licences	m³/year	No. licences	m³/year	No. licences
Alluvium	6 819	1	81 433	3	—	—	—	—	88 252	4
Alluvium/Lias	14 097	6	3 205	3	—	—	—	—	17 302	9
Lias	35 604	34	10 206	5	—	—	—	—	45 810	39
Alluvium/Mercia Mudstone Group	12 561	8	15 456	4	—	—	—	—	28 017	12
Mercia Mudstone Group	205 802	101	118 605	10	—	—	121 651	2	446 058	113
Sherwood Sandstone Group	6 937	1	—	—	—	—	—	—	6 937	1
Silurian	—	—	—	—	—	—	—	—	—	—
Old Red Sandstone	755	12	—	—	—	—	—	—	755	12
Malverns Complex	—	—	—	—	43 983	2	—	—	43 983	2
Totals	282 575	163	228 905	25	43 983	2	121 651	2	677 114	192

Data provided by the National Rivers Authority, Severn-Trent Region

Waters obtained from various mineral springs in the Malvern area have considerable historical and current importance.

Precambrian

The igneous rocks of the Malverns Complex possess extremely low primary porosity and permeability, and groundwater storage and movement is restricted to fissure flow in the fractures. Due to the relatively small outcrop area of the Malvern Hills and nature of the aquifer, spring discharge varies greatly, falling to a minimum during prolonged dry periods when recharge is absent.

Numerous springs (often referred to locally as 'wells') are located along the margins of the outcrop where the rocks are faulted against or overlain unconformably by less permeable strata — to the west Lower Palaeozoic, to the east, the Mercia Mudstone Group (e.g. Prime's Well [7600 4030]). Some springs also issue from fissures in the Malverns Complex and are reputed to be of particular purity and softness (e.g. St Anne's Well [7715 4578] and Holy Well [7706 4231]). The quality of water from these sources is generally very good, having a very low dissolved mineral content, whereas waters from springs at the outcrop margins, although also generally of good quality, tend to contain an increased mineral content due to contact with the adjacent argillaceous rocks (Richardson, 1930). A chemical analysis of a water sample from St Anne's Well is shown in Table 17.

Many of the spring waters were considered to possess beneficial medicinal properties and their use, both internally and externally, became very popular during the late 18th and early 19th century. The Malvern area developed and prospered during that period, with a large number of hydropathic establishments being set up before the popularity of the 'Water Cure' gradually waned.

Spring water for public consumption has been bottled in the area for a considerable period. Water obtained from St Anne's Well was bottled commercially by Schweppes during the early 1900s (Richardson, 1930), as was water obtained from Prime's Well [7600 4030] (also referred to as Pewtriss Well by Richardson, 1935). Currently the water from Prime's Well is bottled under the internationally known 'Malvern Water' label. The company is licensed to abstract 109 m^3/day or 39 823 m^3/year for bottling and other domestic purposes. Water is also bottled from Holy Well, but on a much smaller scale, licensed abstraction being only 11.4 m^3/day (4150 m^3/year).

Silurian and Devonian

The Llandovery Series May Hill Sandstone Group contains numerous well-cemented sandstones, whereas the Wenlock, Ludlow and Přídolí strata are mostly argillaceous, apart from the Woolhope, Much Wenlock, Aymestry and Bishop's Frome limestones and the Downton Castle Sandstone. The Devonian St Maughans Formation (Lower Old Red Sandstone) consists of argilla-

ceous strata and well-cemented sandstones. All these strata have low primary porosities and permeabilities and groundwater storage and movement occurs predominantly within and through fissures.

Virtually all of the few wells and boreholes have small yields of generally less than 0.5 litres per second (l/sec). Water quality is generally good but locally hard (Table 17).

The Colwall Borehole [7572 4265] penetrated 20.8 m of Ludlow beds beneath 353.6 m of Raglan Mudstone. Water, not encountered until penetrating the Downton Castle Sandstone (p.36), was a brine which overflowed at the surface at a rate of only 0.5 l/min.

In the Raglan Mudstone Formation, the yield from a well is dependent on the size, number and degree of interconnection of water-bearing fissures penetrated. Yields generally range from about 0.3 to 0.7 l/sec, but dry and very low-yielding boreholes are common. The highest yield recorded in the district, of 1.5 l/s for only 3 m drawdown, was obtained from a 35 m deep, 150 mm diameter borehole [7100 4330] at Bosbury. Water quality is generally good but hard.

Permian, Triassic and Jurassic

SANDSTONES

The Permian Bridgnorth Sandstone Formation and the Triassic Sherwood Sandstone Group (Kidderminster, Wildmoor Sandstone and Bromsgrove Sandstone formations) constitute the only major potential aquifer present in the district. Smith and Burgess (1984) considered the Bromsgrove and Wildmoor sandstones as a single multi-layered aquifer unit, in which lightly cemented fine-grained sandstones have some argillaceous beds. Sandstone porosity near surface ranges between 20 and 30 per cent and intergranular permeability between 0.5 and 5 m per day (m/d). At depth the porosity does not appear to vary greatly but permeability, although still significant, may be rather less than at outcrop. The overall transmissivity of the Wildmoor Sandstone at Kempsey [8609 4933] was estimated to be about 39 m^2/day, that of the Bromsgrove Sandstone about 92 m^2/day, with less than 50 per cent of the total thickness contributing to aquifer performance (Smith and Burgess, 1984).

The Kidderminster Formation appears to possess a porosity of less than 15 per cent at depth due to very poor sorting and a high degree of carbonate cementation. In consequence it is, in the absence of fractures, likely to act as an aquitard between the more permeable units above and below.

The Bridgnorth Sandstone is only lightly cemented and has a porosity near outcrop ranging from 25 to 30 per cent. These values appear to be maintained at depth in the Worcester Basin and total transmissivity is likely to be of the order of 130 m^2/day (Smith and Burgess, 1984). Black and Barker (1981) indicated that recharge to the sandstone aquifer was likely to occur at the basin margins, generating roughly horizontal groundwater flow through the sandstones towards the centre of the

Table 17 Typical chemical analyses of groundwater in the Worcester district.

LOCATION	St Ann's Well Malvern	Knightwick Sanatorium[1] Doddenham	Kempsey[2]				Barbourne Brewery Worcester	Lunn's Well[1] Defford
NATIONAL GRID REFERENCE	7715 4578	735 568	8609 4933				8472 5661	c.930 430
TYPE OF SOURCE	Spring	Well	Borehole				Borehole	Well
AQUIFER	Precambrian Malverns Complex	Silurian May Hill Sandstone Group	Triassic Wildmoor Sandstone		Permian Bridgnorth Sandstone		Triassic Mercia Mudstone Group	Jurassic Lower Lias
DATE OF ANALYSIS	1.5.1989	8.7.1905	1.6.1979 (a)	1.6.1979 (b)	4.7.1979 (c)	4.7.1979 (d)	2.1.1948	4.8.1915
pH	7.15	na	8.05	7.35	6.90	7.30	7.5	na
TOTAL DISSOLVED SOLIDS mg/l	na	660	6010	5580	22260	28850	1100	2540
Calcium (Ca^{2+}) mg/l	4.8	87.3	340	500	1500	2100	186.5	490.1
Magnesium (Mg^{2+}) mg/l	3.2	33.0	90	140	290	450	52.8	53.2
Sodium (Na^+) mg/l	6.1	24.7	1800	1320	6600	8000	62.0	54.9
Potassium (K^+) mg/l	0.6	59.4	80	20	95	160	(62.0)	8.0
Bicarbonate (HCO_3^-) mg/l	16.4	287.1	na	2	na	96	56.4	191.4
Sulphate (SO_4^{2-}) mg/l	16.7	93.5	820	420	1560	1580	489.4	1312.3
Chloride (Cl^-) mg/l	6.0	56.6	2840	3130	12000	16300	75.0	17.6
Nitrate (NO_3^-) mg/l	na	nil	na	na	na	na	138.2	3.0

1   data derived from 'Wells and Springs of Worcestershire' (Richardson, 1930)
2   after Smith and Burgess (1984)
(a)   obtained from a drill stem test between 936 to 942 m below ground level
(b)   centrifuged from a core sample from 936 m below ground level
(c)   obtained from a drill stem test between 1293 to 1299 m below ground level
(d)   centrifuged from a core sample from 1486 m below ground level

na   not analysed

basin where upward vertical flow through the overlying low permeability mudstones occurs. Within the district, the sandstone aquifer crops out in only a few small areas along the western margin of the Worcester Basin. Elsewhere the sandstones are overlain and generally confined by up to 1100 m of mudstones of the Mercia Mudstone Group, a factor which has limited the use of the aquifer. The small extent of the outcrop areas severely limits the potential for direct recharge to the aquifer within the district, most recharge taking place in the area of extensive outcrop in the Droitwich district to the north.

Apart from a few shallow wells only one borehole [7470 5283] near Alfrick is known to have been drilled entirely in the Bromsgrove Sandstone at outcrop. It was 150 mm in diameter, 45.7 m deep, had a rest water level of 5.5 m below ground level (bgl) and yielded 1.1 l/sec, but became disused in 1965 and was subsequently backfilled.

In the Doddenham area a number of springs issue from the sandstone aquifer close to the East Malvern Fault. The most important of these is Nipple Well [7414 5685], the only Sherwood Sandstone Group source from which abstraction is currently licensed (6935 m³/year).

Another spring of local importance is Black's Well [7372 5529] which issues from a thin sandstone to the west of Lulsley [7372 5529]. The quality of water obtained from these outcrop areas is good, with total hardness ranging between 200 and 300 milligrams per litre (mg/l) (as $CaCO_3$) and chloride ion concentrations of about 20 mg/l.

Rest water levels in the area of the sandstone aquifer confined by the Mercia Mudstone Group appear generally to be above or just below ground surface. In the Kempsey Borehole [8609 4933] the artesian head is about 20 m above surface. Three deep, water supply boreholes penetrating the Bromsgrove Sandstone beneath Mercia Mudstone cover have been drilled in the district. A borehole of 203 mm diameter was drilled at Malvern Link [7881 4918] in 1903 in response to severe water shortages in the Malvern area. It had a total depth of 267.6 m, having penetrated the base of the Mercia Mudstone at 214.9 m and yielded 8.9 l/sec for a drawdown of 93 m. This was inadequate and a second borehole of 305 mm diameter and a total depth of 289.6 m was drilled nearby [7885 4923] to provide an emergency supplementary supply. Yields of 4.7 l/sec for 55 m of drawdown and 8.4 l/sec for 85 m of drawdown were obtained during testing. By 1972 the boreholes had become contaminated and were abandoned. A borehole at Upton upon Severn [8343 4041] drilled in 1912 to a depth of 518.2 m penetrated the base of the Mercia Mudstone Group at 396.3 m. The borehole was also artesian and yielded 3.8 l/sec for a drawdown of 10 m. At a later date explosives were apparently used in the borehole, presumably in an attempt to improve the yield, but this resulted in the casing becoming cracked with subsequent contamination of the borehole.

Groundwater chemistry of the confined section of the sandstone aquifer suggests that, although more mineralised than at outcrop, waters from the boreholes at Malvern Link and Upton upon Severn were (prior to contamination) of an adequate quality for public supply. The chemistry of the Kempsey Borehole (Table 17) shows the groundwater to be considerably more saline, and that groundwater from the Permian sandstones is of a much greater salinity and hardness than that from the Triassic sandstones.

MERCIA MUDSTONE GROUP

Although normally considered to be an aquiclude, or at best a very minor aquifer, the Mercia Mudstone Group is an important source of groundwater in the Worcester district. Licensed abstraction, including a few sources where there may be some contribution from overlying alluvium and terrace deposits, amounts to 66 per cent of total licensed abstraction from groundwater sources in the district (Table 16).

Up to 1100 m thick in the district, the group comprises mudstones with thin impersistent siltstones and sandstones (skerries). The Arden Sandstone Formation is a thicker, laterally persistent unit with varying proportions of sandstone. The mudstones are effectively impermeable, with a hydraulic conductivity of the order of only 1×10^2 m/d (Black and Barker, 1981). However, usable supplies of water are contained within and flow through fissures in the sandstones. Rest water levels are commonly within 10 m of surface, occasionally overflowing and only rarely in excess of 20 m bgl. Yields vary greatly, but mainly range from 0.3 to 1.3 l/s, although corresponding drawdowns are commonly large. The largest yield obtained from a borehole drilled entirely in the mudstones was from a 68.9 m deep, 250 mm diameter borehole which yielded 4.4 l/sec for the unusually small drawdown of only 9 m. Natural recharge of groundwater in the sandstones is slow due to their enclosure within the mudstones and sustainable yields tend to diminish as groundwater in storage is depleted by pumping.

Groundwater quality is generally good, although very hard, often exceeding 500 mg/l (as $CaCO_3$). Chloride ion concentrations are normally less than 50 mg/l and only rarely exceed 200 mg/l. High sulphate (due to the dissolution of gypsum) and iron concentrations occur locally, and hydrogen sulphide odour has only been reported from wells at Blackfield Farm, Tinkers' Cross [781 565] and Upper Gurnox Farm, Doddenham [749 574] (Richardson, 1930).

The analysis of a water sample from the Mercia Mudstone given in Table 17 is typical, apart from the high sulphate concentration.

PENARTH GROUP AND LIAS GROUP

The mudstones, limestones and sandstones of these groups are relatively impermeable, but capable of yielding small quantities of water, particularly where fissured limestones are penetrated. Few boreholes have been drilled into the strata in the district, although numerous shallow wells, presumably now disused, were reported by Richardson (1930). A number of springs issue from close to the junction between the Penarth and Mercia Mudstone groups as well as from Lias limestones, particularly in the north-east of the district.

Rest water levels are commonly within 10 m of surface but yields are almost uniformly low, generally being less than 0.5 l/sec. Yields are also likely to decline with time due to slow rates of natural recharge and depletion of limited storage due to pumping. Water quality is highly variable even over small areas. Groundwaters are hard, normally exceeding 500 mg/l (as $CaCO_3$) and ranging up to over 4000 mg/l (as $CaCO_3$). Chloride ion concentrations are also highly variable but rarely less than 20 mg/l and frequently in excess of 500 mg/l in more heavily mineralised waters. High sulphate and iron concentrations are common and Richardson (1930) reported an odour of hydrogen sulphide from water obtained from several wells in the Lower Lias Clay. A chemical analysis of a water sample from a well penetrating the Lower Lias is shown in Table 17.

Alluvium and River Terrace Deposits

Extensive alluvial and terrace deposits occur in the valleys of the Severn, Teme and Avon. The terrace sands and gravels are generally too thin to constitute a usable aquifer. A maximum thickness of 10 m was proved in a borehole [8471 4933] in the Second Terrace near Kempsey, and 10.4 m was deduced from resistivity sounding in this terrace to the south [8511 4603]. Records show boreholes to be commonly dry or the deposits to be saturated for only a small part of their thickness. In consequence, many boreholes and wells continue into the underlying strata. A number of springs issue from the base of terrace deposits.

Sustainable yields of about 1 l/s may be expected from boreholes where a sufficient saturated thickness exists. A borehole [8529 4721] yielded 4.5 l/s on test pumping and another [8592 4756] yielded 0.1 l/s, with a flow that persisted in dry spells when other wells were dry. Rest water levels are within a few metres of ground level. It appears that yields from boreholes penetrating a relatively thin saturated thickness of sand and gravel overlying mudstones are commonly much higher than that which would be expected solely from the mudstones.

For boreholes and wells penetrating only alluvial and terrace deposits total hardness generally ranges up to about 450 mg/l (as $CaCO_3$) whilst chloride ion concentrations rarely exceed 50 mg/l.

MINERAL RESOURCES

Igneous rock

The hard, tough rocks of the Malverns Complex provided a source of raw material for use as roadstone chippings and setts in an area virtually devoid of comparable materials. They were also used extensively in the Malvern area for building stone (Bennett, 1942).

At the end of the last century, when official lists of quarries first began to be published systematically, there were 13 working quarries in the Malverns. Since then, the industry gradually became concentrated into a few large quarries before eventually becoming extinct. In 1934 the Pyx Granite Company worked the quarries at North Malvern (North Hill, Scar and Tank), while the Malvern Hill Granite Company worked Tolgate Quarry and Upper Wyche Quarry. The only other operations were at Hollybush and Gullet, to the south of the district.

In the late 1930s the Malvern Hill Granite Company opened Earnslaw Quarry, which developed into a large operation, but continuing postwar decline resulted in further reductions in activity. Tank (Pyx) Quarry, the last remaining operation in the district, then owned by the Amalgamated Roadstone Group, closed in 1969. The early 1980s saw the end of quarrying throughout the Malvern Hills when Hollybush and Gullet quarries ceased to operate.

Technical data supplied by the successors to the last operating company (ARC) indicate that the product of Tank Quarry was a quartz diorite which provided a good aggregate, well suited for coating with bitumen and use in road surfacing. Properties reported by the operating company in the final years of activity are as follows:

Specific gravity	2.72
Apparent specific gravity	2.76
Water absorption	0.6%
Aggregate crushing value	16.7
Aggregate impact value	14.8
Aggregate abrasion value	4.4
Polished stone value	57

The quarry and its method of operation have been described in detail in the trade literature (Anon., 1964).

In view of the high visibility of quarrying operations in the Malverns and their designation as an Area of Outstanding Natural Beauty, it seems unlikely that further quarrying will take place in the foreseeable future.

Limestone

Silurian limestones have been exploited in numerous small workings throughout the district, although only those in the Much Wenlock Limestone have been of any real importance during this century. Exploitation never evolved on a large scale and all operations are now closed.

The fairly steep westerly dip of the Much Wenlock Limestone over most of its outcrop, together with its somewhat variable quality, have limited both the access-ible reserves and the demand for the product. The presence of mudstone has an adverse effect on the physical properties of the limestone when used as an aggregate and the material is probably now only suitable for fill and hardcore. However, at one time it was worked for agricultural purposes and roadstone.

The largest quarry in the Much Wenlock Limestone was at Whitman's Hill, Storridge [748 483], which ceased to operate in 1984. Another quarry at Blackhouse Lane, Suckley was still operating in the postwar period, but other recorded quarrying at Collins Green ceased before the First World War and Tundridge Quarry near Suckley and Park Quarry near Colwall closed during the 1920s. A quarry referred to in the records as 'near Ankerdine Hill' is probably the one at Knightwick [7370 5592] in the Aymestry Limestone, which was operating until 1912.

The calcrete limestones of the Old Red Sandstone were formerly dug on a small scale for calcining for agricultural lime and for mortar and aggregate for local use. The Bishop's Frome Limestone was the best limestone and the most exploited, including, it is said, in former underground workings at Bishop's Frome [6642 4870], close to the west of the district. Working was recorded in the limestone in the Lower Old Red Sandstone at Birchend [c.670 447], Castle Frome during the final years of the 19th century.

Jurassic limestones were worked in the 19th century for lime making and building stone. The Wilmcote Limestone and the Inferior Oolite were used as building stones; the Wilmcote and the Rugby Limestone were exploited for lime-making. There is a possibility of renewed interest in working these limestones for building stone.

Sandstone

Many of the sandstones of the district, including those of the Lower Old Red Sandstone, Arden Sandstone and Bromsgrove Sandstone, have been quarried in the past for local building use. There is a recorded working in the basal sandstones of the Raglan Mudstone Formation at Wellington Heath in 1896. In the late 19th century and early years of the present one the St Maughans Formation, Lower Old Red Sandstone, was quarried for building stone at two quarries [7156 4760; 7187 4750] near Ridgeway Cross. The second one closed during the First World War and a quarry on Bromyard Downs [? 668 554] closed in the previous decade.

Common clay

There is, at present, no clay working within the district, although in the past extensive use has been made of the Mercia Mudstone (including the Blue Anchor Formation) for brickmaking, and there has been some exploitation of Silurian, Devonian and Jurassic mudstones and alluvial clays.

At Worcester the Mercia Mudstone was worked at Gregory's Bank from the 19th century until the late 1950s for brickmaking and at Norton near Worcester until the late 1960s. At Malvern the Belmont Brick

Works also exploited Mercia Mudstone from the last century until 1950 (Plate 8). The Mercia Mudstone was also widely dug in shallow pits for agricultural marl in the last century.

The only other pit of any significance was the Linton Tile Works Stream Hall Quarry [6671 5388] (Plate 6), which exploited mudstones of the St Maughans Formation until 1981.

Sand and gravel

At present (1991) sand and gravel extraction is confined to two operations. One is at The Brays Pit [731 443] in Mathon Sand and Gravel, the other is in terrace gravels at Upton upon Severn. Resources of sand and gravel are extensive in the valleys of the Severn, Teme and Avon.

Gravels of the Severn Valley and Teme Valley

Assessment of the deposits in most of the Severn Valley that lies within the district has been carried out by BGS, including some resistivity surveys (Ambrose et al., 1985, 1987; Moorlock et al., 1985). Details of former workings in the area around Worcester are given by Wills (1938) and Richardson (1956).

The terrace gravels of the Severn and Teme valleys comprise the major resource of sand and gravel in the district. Although there are six terraces, only the lower ones, particularly the second and third terraces, are important as sand and gravel resources. The floodplain of the Severn is covered by a layer of silt or silty clay which averages 5 to 6 m in thickness. Much of it is underlain by sand and gravel which may be workable in places, particularly north of Worcester where the gravels lie at the surface of the First Terrace. The terrace sands are composed of quartz grains whereas the gravels are mainly of 'Bunter' quartzite and vein quartz pebbles, together with pebbles of Malverns Complex, Triassic sandstone, flint and coal. The gravels are being worked at present at Upton upon Severn; extraction has taken place within the last 40 years at other sites near Worcester. The current workings at Upton produce gravel, fine aggregate for concrete and building sand. More sand is produced than gravel. The gravel is reported to comprise 48 per cent sandstone, 17 per cent quartzite, 15 per cent limestone, 9 per cent igneous rock, 8 per cent chert and 3 per cent ironstone.

The gravels of the Teme Valley remain unexploited apart from minor local pitting.

Gravels of the Avon Valley

The Avon Valley has five gravel terraces. The upper ones have a high clay content and are thin, whereas in the First and Second terraces the gravel is up to 5 m thick in places. The gravels comprise 'Bunter' pebbles, and pebbles of Jurassic limestone, flint and rolled Liassic fossils. Gravel deposits are also present in the valley of the Bow Brook, a tributary of the Avon. There are recently abandoned workings in the Avon Second Terrace deposits at Fladbury near Pershore, just outside the district to the east.

Mathon Sand and Gravel

There are significant deposits of sand and gravel south of Mathon in the Cradley valley, west of the Malverns, where they lie buried below glaciolacustrine, glacial and gelifluction deposits. To the north, small deposits have been worked to a minor extent in the past (Barclay et al., 1990). The deposits currently being worked at The Brays Pit by Western Aggregates Ltd are up to 9 m thick and comprise basal gravels (about 5 m thick) overlain by sands. The gravels have a pebble suite consisting of local and exotic lithologies, the latter including 'Bunter' pebbles (Barclay et al., 1992). The material is used mainly for local use in ready-mix concrete, but is sometimes carried up to 50 miles from the site.

Warner's Farm Pit [7398 4525], operated by Moreton C Cullimore (Gravels) Ltd, ceased extraction in 1990 and is currently importing sand from outside the district for washing. A fine-grained, clean sand, lying between basal and upper gravels, was dug for tile making, as well as for local use as a building sand.

Head

Solifluction, gelifluction and colluvial deposits, collectively termed 'Head', occurring on the east side of the Malverns consist mainly of Malverns Complex rock debris, together with some 'Bunter' pebbles, in a clayey matrix. Although they form extensive sheets in the area between the Malverns and the River Severn they are rarely more than 1 m thick, are of very limited importance as a gravel resource, and have been dug only to a minor extent. Similar deposits of fan gravels on the flanks of Bredon Hill consist of limestone debris and sand (Whittaker, 1972b).

Coal

Phillips (1848) noted that the coals of the Highley Formation had been worked 30 years prior to his survey at Berrow Hill (Chapter 6). They have remained unworked since, are probably thin, of poor quality and of no economic signicance. Variscan inversion and erosion of the Lower Palaeozoic cover in the region of the Worcester Basin removed all but remnants of the Coal Measures within the district, although they are present to the east of the Inkberrow Fault, below the Mesozoic cover (Chapter 10).

Hydrocarbons

The two deep boreholes drilled within or close to the district — Kempsey and Collington — were devoid of hydrocarbons. The Kempsey Borehole [86090 49334] showed that oil prospectivity in the Worcester Basin within the district is low, with the absence of any Lower Palaeozoic rocks to provide a hydrocarbon source. The Collington Borehole [6460 6100], north-west of the district, proved the absence of Cambrian/Ordovician rocks which might have been a possible hydrocarbon source. Seismic reflection traverses show that this is the case west of the northern Malverns, although Cambrian/Ordovician rocks wedge in to the west of Collington (Chapter 10). They are also present west of the southern Malverns, and any future oil interest is likely to focus on structures in this area which may, in addition to having Cambrian/Ordovician source rocks at depth, also have thick Llandovery sandstones to provide a possible reservoir.

Metalliferous minerals

Anomalously high values of gold (4.5 ppm) and silver (up to 64 ppm) have been reported from late-stage reddened granite in the Malverns Complex (Brammall and Dowie, 1936). However, geochemical studies by Bullard (1975) in the north Malverns indicate that neither occurs in economically viable amounts, and the fact that none of 37 granites (including 60 per cent late granites) analysed by Bullard contains detectable quantities of gold casts doubt on Brammall and Dowie's results. The highest level of gold recorded by Bullard (0.5 ppm) occurs in a diorite on the western slope of Worcestershire Beacon, that of silver (4.05 ppm) is from a chalcosine vein at Pyx Quarry.

Other metals occur mainly as traces in the Malverns Complex, although Bullard (1975) recorded over 300 ppm of lead and over 200 ppm of zinc around the locality known as Gold Mine [768 441] north of Wyche, over 250 ppm of lead, over 800 ppm of zinc and over 130 ppm of copper on the col [768 456] between Worcestershire Beacon and Sugarloaf Hill, and over 190 ppm of copper at Earnslaw Quarry [770 444].

Evaporite minerals

Gypsum and anhydrite occur as nodules, seams and veins in the Mercia Mudstone Group, but no economically viable deposits are known. A few collapse hollows in the north-east of the district indicate that the Droitwich Halite is present at depth, where its east–west extent is probably confined by the Smite and Inkberrow faults (Chapter 7). The southern extent of the deposits is unknown, but they may be present at depth across the district to the east of the Smite–Pirton–Tewkesbury fault system, and strong seismic reflectors within the Eldersfield Mudstone may be halite beds (Chapter 10).

Geothermal resources

The geothermal resources of the groundwater of the Worcester Basin have been investigated by the BGS (Smith, 1986; Smith and Burgess, 1984). The Bridgnorth Sandstone provides the greatest potential in the north of the basin, the Wildmoor Sandstone and Bromsgrove Sandstone in the south. Temperatures range between 20°C and 60°C in the sandstones, with a mean of about 45°C, decreasing to 20°C to 25°C at the base of the Mercia Mudstone Group. Heat flow values of 34 milliwatts per square metre (mW/m^2) and 41 mW/m^2 were measured in Malvern Link [788 492]

and Worcester Heat Flow [862 576] boreholes repectively.

Hot rock geothermal resources of the Malverns Complex were investigated by drilling a borehole at Wyche [770 441] (Gebsky et al., 1987). The undisturbed heat flow is probably less than 61mW/m^2 and comparable with values for other Precambrian rocks of the Midlands microcraton.

POTENTIAL GEOLOGICAL RISK FACTORS

This brief account provides a summary of the geological factors that should be considered in planning development in the district and those which pose a potential pollution risk to groundwater resources. The geological map provides a first indication of ground conditions by showing the extent of deposits, such as landslip, soft clays and peat, which present an obvious risk to structures. It should be stressed that BGS maps, even at 1:10 000 scale, do not provide sufficiently detailed information for site investigation purposes, and that detailed, site-specific investigation is necessary prior to any development.

Made Ground

There are no landfill sites currently operative within the district. Of the 20 recorded sites, seven were licensed to accept only soil and rubble and should present no threat to groundwater quality. Two sites which accepted industrial waste, Bransford Court, Rushwick [808 524] and Masonic Hall, Rainbow Hill [859 555], were both on the Mercia Mudstone Group and therefore represent little risk to the overall groundwater resource, although local pollution plumes are possible. The remaining sites were all used for domestic waste, eight being located on the Mercia Mudstone Group, and one each on the Lias Group, Malverns Complex (Tank Quarry, North Malvern [768 470]) and Old Red Sandstone (Warren Wood, Brockhampton [675 548]). The sites on the mudstone strata pose little risk to water quality, whilst those on the Malverns Complex and Old Red Sandstone, having been closed, appear to represent a declining risk.

Methane

The decomposition of domestic landfill generates methane gas which may accumulate and migrate in potentially explosive quantities if the landfill site has not been designed to allow venting. Any proposed development on areas of Made Ground should therefore be preceded by an investigation to ascertain the nature of the fill and the risk of methane accumulation.

Landslip

A summary of the landslip deposits of the district is given on p.93. Slopes on the Silurian Wyche Formation and the Jurassic Lower and Upper Lias Clay formations are particularly prone to slipping, and large areas of unstable or potentially unstable ground are present on the flanks of Ankerdine Hill and Bredon Hill. There are also extensive areas of slipping in the Lias outcrop on the banks of the River Avon.

Head

Head deposits (p.93) may present engineering problems when excavated on steep slopes, particularly on the very steep slopes of the Malvern Hills.

Soft ground

Included under this heading are the alluvial deposits of the district which contain compressible clays, silts and (locally) peat. Sands within these deposits may be thixotropic below the water table. Similarly, the till deposits of the district, and any included or underlying sands may present foundation or excavation difficulties. Soft ground may also be encountered on slopes underlain by strata comprising permeable and impermeable layers, where springs emerge from the junction of these layers.

REFERENCES

Most of the references listed below are held in the library of the British Geological Survey at Keyworth, Nottingham. Copies can be purchased, subject to the current copyright legislation.

ACLAND, H D. 1898. On a Volcanic Series in the Malvern Hills near the Herefordshire Beacon. *Quarterly Journal of the Geological Society of London*, Vol. 54, 556–563.

AINSWORTH, R B. 1991. Discussion on palynofacies in a Late Silurian regressive sequence in the Welsh Borderland and Wales. *Journal of the Geological Society of London*, Vol. 148, 781–784.

ALDRIDGE, R J. 1972. Llandovery conodonts from the Welsh Borderland. *Bulletin of the British Museum of Natural History (Geology)*, Vol. 22, 125–231.

ALDRIDGE, R J, and SCHÖNLAUB, H P. 1989. Conodonts. 274–279 *in* A global standard for the Silurian System. HOLLAND, C H and BASSETT, M G (editors). *National Museum of Wales, Geological Series*, No. 9.

ALLEN, J R L. 1964. Studies in fluviatile sedimentation: six cyclothems from the Lower Old Red Sandstone, Anglo-Welsh Basin. *Sedimentology*, Vol. 3, 163–198.

ALLEN, J R L. 1974a. Sedimentology of the Old Red Sandstone (Siluro-Devonian) in the Clee Hills area, Shropshire, England. *Sedimentary Geology*, Vol. 12, 73–167.

ALLEN, J R L. 1974b. Source rocks of the Lower Old Red Sandstone: exotic pebbles from the Brownstones, Ross-on-Wye, Hereford and Worcester. *Proceedings of the Geologists' Association*, Vol. 85, 493–510.

ALLEN, J R L. 1985. Marine to fresh water: the sedimentology of the interrupted environmental transition (Ludlow-Siegenian) in the Anglo-Welsh region. *Philosophical Transactions of the Royal Society of London*, Vol. B309, 85–104.

ALLEN, J R L, and CROWLEY, S F. 1983. Lower Old Red Sandstone fluvial dispersal systems in the British Isles. *Transactions of the Royal Society of Edinburgh: Earth Sciences*, Vol. 74, 61–68.

ALLEN, J R L, and WILLIAMS, B P J. 1979. Interfluvial drainage on Siluro-Devonian alluvial plains in Wales and the Welsh Borders. *Journal of the Geological Society of London*, Vol. 136, 361–366.

ALLEN, J R L, and WILLIAMS, B P J. 1981a. Sedimentology and stratigraphy of the Townsend Tuff Bed (Lower Old Red Sandstone) in South Wales and the Welsh Borders. *Journal of the Geological Society of London*, Vol. 138, 15–29.

ALLEN, J R L, and WILLIAMS, B P J. 1981b. *Beaconites antarcticus*: a giant channel-associated trace fossil from the Lower Old Red Sandstone of South Wales and the Welsh Borders. *Geological Journal*, Vol. 16, 255–269.

AMBROSE, K. 1984. *Geological notes and local details for 1:10 000 sheets: SO96SE and 95NE (N) (Hanbury and Stock Green)*. (Keyworth: British Geological Survey.)

AMBROSE, K. In preparation. The stratigraphy of the Mercia Mudstone Group of the Worcester Basin.

AMBROSE, K, CANNELL, B, and MOORLOCK, B S P. 1987. The mapping and assessment of aggregate resources in the south Midlands and Welsh Borderland. 347–353 *in Planning and engineering geology*. BELL, M G, CRIPPS, J C, and O'HARA, M (editors). *Geological Society Engineering Geology Special Publication*, No. 4.

AMBROSE, K, MOORLOCK, B S P, and CANNELL, B. 1985. Geology of Sheet SO84. (Keyworth: British Geological Survey.)

ANON. 1964. Industrial television at Pyx Granite. *Mine and Quarry Engineering*, Vol. 30, 190–197.

ARKELL, W J. 1933. *The Jurassic System in Great Britain*. (Oxford: Clarendon Press.)

ARTHURTON, R A. 1980. Rhythmic sedimentary sequences in the Triassic Keuper Marl (Mercia Mudstone Group) of Cheshire, northwest England. *Geological Journal*, Vol. 15, 43–58.

AUDLEY-CHARLES, M G. 1970a. Stratigraphical correlation of the Triassic rocks of the British Isles. *Quarterly Journal of the Geological Society of London*, Vol. 126, 19–47.

AUDLEY-CHARLES, M G. 1970b. Triassic palaeogeography of the British Isles. *Quarterly Journal of the Geological Society of London*, Vol. 126, 48–89.

BARCLAY, W J, BRANDON, A, ELLISON, R A, and MOORLOCK, B S P. 1990. Details of the Pleistocene deposits in the Malvern Hills area. *British Geological Survey Technical Report*, WA/90/92.

BARCLAY, W J, BRANDON, A, ELLISON, R A, and MOORLOCK, B S P. 1992. A Middle Pleistocene palaeovalley-fill west of the Malvern Hills. *Journal of the Geological Society of London*, Vol. 149, 75–92.

BARCLAY, W J, RATHBONE, P A, WHITE, D E, and RICHARDSON, J B. 1994. Brackish water faunas from the St Maughans Formation: the Old Red Sandstone section at Ammons Hill, Hereford and Worcester, UK re-examined. *Geological Journal*, Vol. 29, 369–379.

BARKER, P. 1969. The origins of Worcester. *Transactions of the Worcestershire Archaeological Society*, Third Series, Vol. 2.

BASSETT, M G. 1974. Review of the stratigraphy of the Wenlock Series in the Welsh Borderland and South Wales. *Palaeontology*, Vol. 17, 745–777.

BASSETT, M G, BLUCK, B J, CAVE, R, HOLLAND, C H, and LAWSON, J D. 1992. Silurian. 37–55 in Atlas of palaeogeography and lithofacies. COPE, J C W., INGHAM, J K, and RAWSON, P F (editors). *Memoir of the Geological Society of London*, No. 13.

BASSETT, M G, COCKS, L R M, HOLLAND, C H, RICKARDS, R B, and WARREN, P T. 1975. The type Wenlock Series. *Report of the Institute of Geological Sciences*, No. 75/13, 1–20.

BATHURST, R G C. 1987. Diagenetically enhanced bedding in argillaceous platform limestones: stratified cementation and selective compaction. *Sedimentology*, Vol. 34, 749–778.

BECKINSALE, R D, THORPE, R S, PANKHURST, R J, and EVANS, J A. 1981. Rb-Sr whole-rock isochron evidence for the age of the Malvern Hills Complex. *Journal of the Geological Society of London*, Vol. 138, 69–73.

BENNETT, A G. 1942. *The geology of Malvernia.* Malvern Naturalists' Field Club.

BENTON, M J, WARRINGTON, G, NEWELL, A J, and SPENCER, P S. 1994. A review of the British Mid Triassic tetrapod assemblages. 131–160 in *In the shadow of the Dinosaurs: Early Mesozoic Tetrapods.* FRASER, N C, and SUES, H-D (editors). (Cambridge University Press.)

BLACK, J H, and BARKER, J A. 1981. Hydrogeological reconnaissance study: Worcester Basin. *Report of the Environmental Protection Unit, Institute of Geological Sciences,* ENPU 81-3.

BLUCK, B J, COPE, J C W, and SCRUTTON, C T. 1992. Devonian. 57–66 in Atlas of palaeogeography and lithofacies. COPE, J C W, INGHAM, J K, and RAWSON, P F (editors). *Memoir of the Geological Society of London,* No. 13.

BLYTH, F G H. 1952. Malvern tectonics — a contribution. *Geological Magazine,* Vol. 89, 185–194.

BLYTH, F G H. and LAMBERT, R St J. 1970. Chemical data from the Malvernian of the Malvern Hills, Herefordshire. *Quarterly Journal of the Geological Society of London,* Vol. 125, 543–555.

BOSWELL, P G H. 1924. The petrography of the sands of the Lias–Inferior Oolite of the west of England. *Geological Magazine,* Vol. 61, 246–264.

BOULTON, W S. 1917. Mammalian remains in the glacial gravels at Stourbridge. *Proceedings of the Birmingham Natural History and Philosophical Society,* Vol. 14, 107–112.

BOWEN, D Q, HUGHES, S, SYKES, G A, and MILLER, G H. 1989. Land-sea correlations in the Pleistocene based on isoleucine epimerization in non-marine molluscs. *Nature, London,* Vol. 340, 49–51.

BOWEN, D Q, ROSE, J, MCCABE, A M, and SUTHERLAND, D G. 1986. Correlation of Quaternary glaciations in England. *Quaternary Science Review,* Vol. 5, 299–340.

BRADSHAW, M J, COPE, J C W, CRIPPS, D W, DONOVAN, D T, HOWARTH, M K, RAWSON, P F, WEST, I M, and WIMBLEDON, W A. 1992. Jurassic. 107–129 in Atlas of palaeogeography and lithofacies. COPE, J C W, INGHAM, J K, and RAWSON, P F (editors). *Memoir of the Geological Society of London,* 18.

BRAMMALL, A. 1940. 52–62 in Report of Easter Field Meeting at Hereford. *Proceedings of the Geologists' Association,* Vol. 51.

BRAMMALL, A, and DOWIE, D L. 1936. The distribution of gold and silver in the crystalline rocks of the Malvern Hills. *Mineralogical Magazine,* Vol. 24, 260–264.

BRANDON, A. 1989. Geology of the country between Hereford and Leominster. *Memoir of the British Geological Survey,* Sheet 198 (England and Wales).

BRASIER, M D, INGHAM, J K, and RUSHTON, A W A. 1992. Cambrian. 13–18 in Atlas of palaeogeography and lithofacies. COPE, J C W, INGHAM, J K, and RAWSON, P F (editors). *Memoir of the Geological Society of London,* No. 18.

BRIDGLAND, D R, KEEN, D H, and MADDY, D. 1986. A reinvestigation of the Bushley Green terrace of the River Severn at the type site. *Quaternary Newsletter,* Vol. 50, 1–6.

BRIDGLAND, D R, KEEN, D H, and MADDY, D. 1989. The Avon terraces: Cropthorne, Ailstone and Eckington. 51–67 in *The Pleistocene of the West Midlands.* KEEN, D H (editor). (Quaternary Research Association, Cambridge.)

BRODIE, P B. 1845. *A history of the fossil insects in the secondary rocks of England.* (London: John Van Voorst.)

BRODIE, P B. 1865. On the Lias outliers at Knowle and Wootton Wawen in south Warwickshire, and on the presence of the Lias or Rhaetic Bone-bed at Copt Heath, its furthest extension hitherto recognised in that county. *Quarterly Journal of the Geological Society of London,* Vol. 21, 159–161.

BRODIE, P B. 1868. A sketch of the Lias generally in England, and of the 'Insect and Saurian Beds', especially in the lower division in the counties of Warwick, Worcester and Gloucester, with a particular account of the fossils which characterise them. *Proceedings of the Warwickshire Naturalists' and Archaeological Field Club,* 1–24.

BRODIE, P B. 1870. On the geology of Warwickshire. *34th Annual Report of the Warwickshire Natural History Society,* 10–34.

BRODIE, P B. 1874. Notes on a railway section of the Lower Lias and Rhaetics between Stratford-on-Avon and Fenny Compton, on the occurrence of the Rhaetics near Kineton, and the Insect Beds near Knowle in Warwickshire, and on the recent discovery of the Rhaetics near Leicester. *Quarterly Journal of the Geological Society of London,* Vol. 30, 746–749.

BRODIE, P B. 1875. The Lower Lias at Ettington and Kineton and on the Rhaetics in that neighbourhood and their furthest extension in Leicestershire, Nottinghamshire, Lincolnshire, Yorkshire and Cumberland. *Warwickshire Natural History and Archaeological Society Annual Report,* No. 39, 6–17.

BROOKS, M. 1968. The geological results of gravity and magnetic surveys in the Malvern Hills and adjacent districts. *Geological Journal,* Vol. 6, 13–30.

BROOKS, M. 1970. Pre-Llandovery tectonism and the Malvern structure. *Proceedings of the Geologists' Association,* Vol. 81, 249–268.

BROOKS, M, and BASSETT, M G. 1970. New exposures of Cambrian and Llandoverian rocks in the southern Malvern Hills. *Proceedings of the Geological Society of London,* No. 1660, 361–363.

BROOKS, M, and DRUCE, E C. 1965. A Llandovery conglomeratic limestone in Gullet quarry, Malvern Hills, and its conodont fauna. *Geological Magazine,* Vol. 102, 370–382.

BROWN, A G. 1983. Floodplain deposits and accelerated sedimentation in the lower Severn basin. 375-397 in *Background to palaeohydrology.* GREGORY, K J (editor). (Chichester: Wiley.)

BRUGMAN, W A. 1986. Late Scythian and Middle Triassic palynostratigraphy in the Alpine realm. *Albertiana,* Vol. 5, 19–20.

BUCKMAN, J. 1850. On some fossil plants from the Lower Lias. *Quarterly Journal of the Geological Society of London,* Vol. 6, 413–418.

BUCKMAN, S S. 1903. The Toarcian of Bredon Hill and a comparison with deposits elsewhere. *Quarterly Journal of the Geological Society of London,* Vol. 59, 445–458.

BULLARD, D W. 1974. Western boundary of the Malvernian, North Malvern Hills, Worcestershire. *Mercian Geologist,* Vol. 5, 65–70.

BULLARD, D W. 1975. Aspects of the geology and geochemistry of the soils and Precambrian rocks of the Malvern Hills, Worcestershire. Unpublished PhD thesis, University of Nottingham.

BULLARD, D W. 1989. *Malvern Hills. A student's guide to the geology of the Malverns.* (Nature Conservancy Council.)

BUTCHER, N E. 1962. The tectonic structure of the Malvern Hills. *Proceedings of the Geologists' Association,* Vol. 73, 103–123.

CALEF, C E, and HANCOCK, N J. 1974. Wenlock and Ludlow marine communities in Wales and the Welsh Borderland. *Palaeontology,* Vol. 17, 779–810.

CALLAWAY, C. 1880. On a second Precambrian group in the Malvern Hills. *Quarterly Journal of the Geological Society of London*, Vol. 36, 536–539.

CALLAWAY, C. 1889. On the production of secondary minerals at shear-zones in the crystalline rocks of the Malvern Hills. *Quarterly Journal of the Geological Society of London*, Vol. 45, 475–503.

CAVE, R. 1977. Geology of the Malmesbury district. *Memoir of the Geological Survey of Great Britain*, Sheet 251 (England and Wales).

CHADWICK, R A. 1985. Seismic reflection investigations into the stratigraphy and structural evolution of the Worcester Basin. *Journal of the Geological Society of London*, Vol. 142, 187–202.

CHADWICK, R A, and SMITH, N J P. 1988. Evidence of negative structural inversion beneath central England from new seismic reflection data. *Journal of the Geological Society of London*, Vol. 145, 519–522.

CHADWICK, R A, KIRBY, G A, and BAILY, H E. 1994. The post-Triassic structural evolution of north-west England and adjacent parts of the East Irish Sea. *Proceedings of the Yorkshire Geological Society*, Vol. 50, 91–102.

CHERNS, L. 1980. Hardgrounds in the Lower Leintwardine Beds (Silurian) of the Welsh Borderland. *Geological Magazine*, Vol. 117, 311–325.

CLARKE, R F A. 1965. Keuper miospores from Worcestershire, England. *Palaeontology*, Vol. 8, 294–321.

COCKS, L R M. 1989. The Llandovery Series in the Llandovery area. 36–50 *in* A global standard for the Silurian System. HOLLAND, C H, and BASSETT, M G (editors). *National Museum of Wales, Cardiff, Geological Series*, No.9.

COCKS, L R M, HOLLAND, C H, RICKARDS, R B, and STRACHAN, I. 1971. A correlation of the Silurian rocks in the British Isles. *Journal of the Geological Society of London*, Vol. 127, 103–136 [also as *Special Report of the Geological Society of London*, No. 1, 1–34].

COCKS, L R M, HOLLAND, C H, and RICKARDS, R B. 1992. A revised correlation of Silurian rocks in the British Isles. *Special Report of the Geological Society of London*, No. 21.

COCKS, L R M, WOODCOCK, N H, RICKARDS, R B, TEMPLE, J T, and LANE, P D. 1984. The Llandovery Series of the type area. *Bulletin of the British Museum of Natural History (Geology)*, Vol. 38, 131–182.

COOK, A H, and THIRLAWAY, I S. 1955. The geological results of measurements of gravity in the Welsh borders. *Quarterly Journal of the Geological Society of London*, Vol. 111, 47–70.

COOPE, G R, SHOTTON, F W, and Strachan, I. 1961. A late Pleistocene fauna and flora from Upton Warren, Worcestershire. *Philosophical Transactions of the Royal Society*, Series B, Vol. 244, 379–421.

COPE, J C W, GETTY, T A, HOWARTH, M K, MORTON, N, and TORRENS, H S. 1980a. A correlation of Jurassic rocks in the British Isles. Part One: Introduction and Lower Jurassic. *Special Report of the Geological Society of London*, No. 14.

COPE, J C W, DUFF, K L, PARSONS, C F, TORRENS, H S, WIMBLEDON, W A, and WRIGHT, J K. 1980b. A correlation of Jurassic rocks in the British Isles. Part Two: Middle and Upper Jurassic. *Special Report of the Geological Society of London*, No. 15.

CORNWELL, J D. 1992. Geophysical investigations in the Worcester district. *British Geological Survey Technical Report*, WK/92/17

COWIE, J W. 1992. Cambrian. 35–61 in *Geology of England and Wales*. DUFF, P McL, D, and SMITH, A J (editors). (The Geological Society of London.)

COWIE, J W, RUSHTON, A W A, and STUBBLEFIELD, C J. 1972. A correlation of Cambrian rocks in the British Isles. *Special Report of the Geological Society of London*, No. 2.

DAVISON, C. 1924. *A history of British earthquakes.* 416 pp. (Cambridge: Cambridge University Press.)

DAWSON, M. 1989. The Severn Valley south of Bridgnorth. 78–100 in *The Pleistocene of the West Midlands: Field Guide*. KEEN, D H (editor). (Quaternary Research Association, Cambridge.)

DE ROUFFIGNAC, C, Bowen, D Q, COOPE, C R, KEEN, D H, LISTER, A M, MADDY, D, ROBINSON, J E, SYKES, G A, and WALKER, M J C. 1995. Late Middle Pleistocene interglacial deposits at Upper Strensham, Worcestershire, England. *Journal of Quaternary Science*, Vol. 10, 15–31.

DEARNLEY, R. 1990. Malvern Hills Igneous Suite. *British Geological Survey Technical Report*, WG/90/11.

DEAN, W T, DONOVAN, D T, and HOWARTH, M K. 1961. The Liassic ammonite zones and subzones of the North-west European Province. *Bulletin of the British Museum, Natural History (Geology)*, Vol. 4, 433–505.

DEPARTMENT OF ENERGY. 1978. *UK land well records. Collington No. 1, Hereford.* (London: HMSO.)

DONOVAN, D T, HORTON, A, and IVIMEY-COOK, H C. 1979. The transgression of the Lower Lias over the northern flank of the London Platform. *Journal of the Geological Society of London*, Vol. 136, 165–173.

DORNING, K J. 1982. Ludlow stratigraphy at Ludlow, Shropshire. *Geological Magazine*, Vol. 119, 615–618.

DORNING, K J, and BELL, D G. 1987. The Silurian carbonate shelf microflora: acritarch distribution in the Much Wenlock Limestone Formation. 266–287 in *Micropalaeontology of carbonate environments*. HART, M B (editor). (Chichester: Ellis Horwood.)

DUMBLETON, M J, and WEST, G. 1966. Studies of the Keuper Marl: Mineralogy. *Road Research Laboratory (Ministry of Defence), Report*, No. 40.

EARP, J R, and HAINS, B A. 1971. *British regional geology: the Welsh Borderland* (3rd edition). (London: HMSO for Institute of Geological Sciences.)

EDMONDS, E A, POOLE, E G, and WILSON, V. 1965. Geology of the country around Banbury and Edge Hill. *Memoir of the Geological Survey of Great Britain*, Sheet 201 (England and Wales).

ELLIOTT, R E. 1961. The stratigraphy of the Keuper Series in southern Nottinghamshire. *Proceedings of the Yorkshire Geological Society*, Vol. 33, 197–234.

FALCON, N L, and TARRANT, L H. 1951. The gravitational and magnetic exploration of parts of the Mesozoic-covered areas of south-central England. *Quarterly Journal of the Geological Society of London*, Vol. 106, 141–170.

FITCH, F J, MILLER, J A, and THOMPSON, D B. 1966. The palaeogeographical significance of isotope age determination on detrital micas from the Triassic of the Stockport-Macclesfield district, Central England. *Palaeogeography, Palaeoclimatology, Palaeoecology*, Vol. 2, 281–312.

FITCH, F J, MILLER, J A, EVANS, A L, GRASTY, R L, and MENEISY, M Y. 1969. Isotopic age determinations on rocks from Wales and the Welsh Borders. 23–45 in *The Pre-Cambrian and Lower Palaeozoic rocks of Wales*. WOOD, A (editor). (University of Wales Press.)

FLEET, W E. 1925.   The chief heavy detrital minerals in the rocks of the English Midlands.   *Geological Magazine*, Vol. 62, 98–128.

FLEET, W E. 1927.   The heavy minerals of the Keele, Enville, 'Permian' and Lower Triassic rocks of the Midlands, and the correlation of these strata.   *Proceedings of the Geologists' Association*, Vol. 38, 1–48.

FLEET, W E. 1930.   Petrography of the Lower Keuper Sandstone of the Midlands.   *Proceedings of the Birmingham Natural History and Philosophical Society*, Vol. 16, 13–17.

FLETCHER, C J N, DAVIES, J R, WATERS, R A, WILSON, D A, and WOODHALL, D G. In press.   Geology of the country around Llanilar and Rhyader.   *Memoir of the British Geological Survey* (Sheets 178 and 179, England and Wales).

GEBSKY, J S, WHEILDON, J A, and THOMAS-BETTS, A. 1987. Detailed investigation of the UK heat flow field (1984–1987). Unpublished report, Geology Department, Imperial College of Science and Technology, London.

GEORGE, T N. 1937.   The geology of the district around Dunhampstead and Himbleton, Worcestershire.   *Summary of Progress of the Geological Survey of Great Britain for 1935*, part 2, 119–129.

GRAUVOGEL-STAMM, L. 1972.   Révision de cônes mâles du 'Keuper Inférieur' du Worcestershire (Angleterre) attribués à *Masculostrobus willsi* Townrow.   *Palaeonographica*, B, Vol. 140, 1–26. [In French]

GRAY, J W. 1914. The drift deposits of the Malverns, and their supposed glacial origin.   *Proceedings of the Birmingham Natural History and Philosophical Society*, Vol 13, 1–18.

GREEN, A H. 1895.   Notes on some recent sections in the Malvern Hills.   *Quarterly Journal of the Geological Society of London*, Vol. 51, 1–8.

GREEN, G W and MELVILLE, R V. 1956.   The stratigraphy of the Stowell Park Borehole.   *Bulletin of the Geological Survey of Great Britain*, Vol. 11, 1–66.

GREIG, D C, WRIGHT, J E, HAINS, B A, and MITCHELL, G H. 1968. Geology of the country around Church Stretton, Craven Arms, Wenlock Edge and Brown Clee.   *Memoir of the Geological Survey*, Sheet 166 (England and Wales).

GRINDLEY, H E. 1925.   Observations on the Midland Drift as seen in two sandpits at Mathon.   *Transactions of the Woolhope Naturalists' Field Club*, Vol. for 1921–23, 176–177.

GROOM, T T. 1899.   The geological structure of the southern Malvern Hills and of the adjacent district to the west. *Quarterly Journal of the Geological Society of London*, Vol. 55, 129–169.

GROOM, T T. 1900.   On the geological structure of portions of the Malvern and Abberley Hills.   *Quarterly Journal of the Geological Society of London*, Vol. 56, 138–197.

GROOM, T T. 1902.   The sequence of the Cambrian and associated beds of the Malvern Hills.   *Quarterly Journal of the Geological Society of London*, Vol. 58, 89–135.

GROOM, T T. 1910.   The Malvern and Abberley Hills, and the Ledbury District.   698–738 in *Geology in the field (Jubilee Vol. II Geologists' Association)*. MONCKTON, H W, and HERRIES, R S (editors).

GROVE, A T. 1953.   Account of a mudflow on Bredon Hill, Worcestershire. April 1951.   *Proceedings of the Geologists' Association*, Vol. 64, 10–13.

HALLAM, A. 1960.   A sedimentary and faunal study of the Blue Lias of Dorset and Glamorgan.   *Philosophical Transactions of the Royal Society of London*, Series B, Vol. 243, 1–44.

HALLAM, A. 1964.   Origin of the limestone–shale rhythm in the Blue Lias of England. A composite theory.   *Journal of Geology*, Vol. 72, 157–169.

HALLAM, A. 1967.   An environmental study of the Upper Domerian and Lower Toarcian in Great Britain.   *Philosophical Transactions of the Royal Society of London*, Series B, Vol. 252, 393–445.

HARDIE, W G. 1969.   *A guide to the geology of the Malvern Hills and adjacent areas.*   (Worcestershire Education Committee.)

HARDIE, W G. 1974.   Excursion to the Malvern Hills.   *Mercian Geologist*, Vol. 5, 71–74.

HARDING, T P. 1985.   Seismic characteristics and identification of negative flower-structures, positive flower-structures and positive structural inversion.   *Bulletin of the American Association of Petroleum Geologists*, Vol. 69, 582–600.

HARRISON, W J. 1877.   On the Rhaetic section at Dunhampstead cutting, Droitwich, and its connection with the same strata elsewhere.   *Proceedings of the Dudley and Midland Geological and Scientific Society and Field Club*, Vol. III, No. 5, 115–126.

HASSAN, S M. 1964.   A comparative study of the sedimentology of the Upper Mottled Sandstone and the Lower Keuper Sandstone of an area west of Birmingham.   Unpublished MSc thesis, University of Birmingham.

HEY, R W. 1959.   Pleistocene deposits on the west side of the Malvern Hills.   *Geological Magazine*, Vol. 96, 403–417.

HEY, R W. 1963.   The Pleistocene history of the Malvern Hills and adjacent areas.   *Proceedings of the Cotteswold Naturalists' Field Club*, Vol. 33, 185–191.

HILL, P J. 1974.   Stratigraphic palynology of acritarchs from the type area of the Llandovery and the Welsh Borderland. *Review of Palaeobotany and Palynology*, Vol. 18, 11–23.

HOLL, H B. 1863.   On correlation of the several subdivisions of the Inferior Oolite in the middle and south of England. *Quarterly Journal of the Geological Society of London*, Vol. 19, 306–317.

HOLL, H B. 1865.   On the geological structure of the Malvern Hills and adjacent areas.   *Quarterly Journal of the Geological Society of London*, Vol. 21, 72–102.

HOLLAND, C H, LAWSON, J D, and WALMSLEY, V G. 1959.   A revised classification of the Ludlovian succession at Ludlow. *Nature, London*, Vol. 184, 1057–1059.

HOLLAND, C H, LAWSON, J D. 1962.   Ludlovian classification — a reply.   *Geological Magazine*, Vol. 99, 393–398.

HOLLAND, C H, LAWSON, J D. 1963.   The Silurian rocks of the Ludlow district, Shropshire.   *Bulletin of the British Museum (Natural History), Geology*, Vol. 8, 95–171.

HOLLAND, C H, LAWSON, J D, and WHITE, D E. 1980.   Ludlow stages.   *Lethaia*, Vol. 13, 268.

HOLLINGWORTH, S E. 1938.   *Summary of progress of the Geological Survey of Great Britain and the Museum of Practical Geology for the year 1937*, No. 35.

HORNER, L. 1811.   On the mineralogy of the Malvern Hills. *Transactions of the Geological Society of London (Series 1)*, Vol. 1. 281–321.

HORTON, A, and POOLE, E G. 1977.   The lithostratigraphy of three geophysical marker horizons in the Lower Lias of Oxfordshire.   *Bulletin of the Geological Survey of Great Britain*, Vol. 62, 13–24.

HOWARTH, M K. 1992. The ammonite family Hildoceratidae in the Lower Jurassic of Britain. Part 1. *Monograph of the Palaeontographical Society, London.* 106 pp.

HULL, E. 1869. The Triassic and Permian rocks of the Midland counties of England. *Memoir of the Geological Survey of Great Britain.*

HURST, J M. 1975a. Wenlock carbonate, level bottom, brachiopod-dominated communities from Wales and the Welsh Borderland. *Palaeogeography, Palaeoclimatology, Palaeoecology,* Vol. 17, 227–255.

HURST, J M. 1975b. The diachronism of the Wenlock Limestone. *Lethaia,* Vol. 8, 301–314.

HURST, J M, HANCOCK, N J, and MCKERROW, W S. 1978. Wenlock stratigraphy and palaeogeography of Wales and the Welsh Borderland. *Proceedings of the Geologists' Association,* Vol. 89, 197–226.

IRVINE, T N, and BARAGER, W R A. 1971. A guide to the chemical classification of the common volcanic rocks. *Canadian Journal of Earth Sciences,* No. 8, 523–548.

JEANS, C V. 1978. The origin of the Triassic clay assemblages of Europe with special reference to the Keuper Marl and Rhaetic of parts of England. *Philosophical Transactions of the Royal Society of London,* Series A, Vol. 289, 549–639.

JONES, R K, BROOKS, M, BASSETT, M G, AUSTIN, R L, and ALDRIDGE, R J. 1969. An Upper Llandovery limestone overlying Hollybush Sandstone (Cambrian) in Hollybush Quarry, Malvern Hills. *Geological Magazine,* Vol. 106, 457–469.

JUDD, J W. 1875. The geology of Rutland and parts of Lincoln, Leicester, Northampton, Huntingdon and Cambridge included in sheet 64 of the one-inch map of the Geological Survey. *Memoirs of the Geological Survey of Great Britain.*

KARPETA, W P. 1990. The morphology of Permian palaeodunes — a reinterpretation of the Bridgnorth Sandstone around Bridgnorth, England, in the light of modern dune studies. *Sedimentary Geology,* Vol. 69, 59–75.

KEAREY, P. 1991. A possible basaltic deep source of the south-central England magnetic anomaly. *Journal of the Geological Society of London,* Vol. 148, 775–780.

KEEN, D H, and Bridgland, D R. 1986. An interglacial fauna from Avon No.3 Terrace at Eckington, Worcestershire. *Proceedings of the Geologists' Association,* Vol. 97, 303–307.

KELLAWAY, G A, and HANCOCK, P L. 1983. Structure of the Bristol district, the Forest of Dean and the Malvern Fault Zone. 88–107 in *The Variscan Fold Belt in the British Isles.* HANCOCK, P L (editor). (Bristol: Adam Hilger Ltd.)

KING, M J, and BENTON, M J. 1996. Dinosaurs in the Early and Mid Triassic? — The footprint evidence from Britain. *Palaeogeography, Palaeoclimatology, Palaeoecology,* Vol. 122, 213–225.

KING, W W. 1934. The Downtonian and Dittonian strata of Great Britain and north-western Europe. *Quarterly Journal of the Geological Society of London,* Vol. 90, 526–570.

KLEMENIC, P. 1987. The geochemistry of Upper Proterozoic lavas from the Red Sea Hills, NE Sudan. 541–552 *in* Geochemistry and mineralisation of Proterozoic volcanic suites. PHARAOH, T C, BECKINSALE, R D, and RICKARD, D (editors). *Special Publication of the Geological Society of London,* No. 33.

LAMBERT, R St J, and HOLLAND, J G. 1971. The petrography and chemistry of the Igneous Complex of the Malvern Hills, England. *Proceedings of the Geologists' Association,* Vol. 82, 323–352.

LAMBERT, R St J, and REX, D C. 1966. Isotopic ages of minerals from the Pre-Cambrian Complex of the Malverns. *Nature, London,* Vol. 209, 605.

LAMONT, A, and GILBERT, D F L. 1945. Upper Llandovery brachiopoda from Coneygore Coppice and Old Storridge Common, near Alfrick, Worcs. *Annals and Magazine of Natural History, 11th Series,* Vol.12, 641–682.

LANKESTER, E R. 1868. Fishes of the Old Red Sandstone of Britain, Part 1, the *Cephalalaspidae. Monograph of the Palaeontographical Society.*

LAPWORTH, C. 1898. Sketch of the geology of the Birmingham district. *Proceedings of the Geologists' Association,* Vol. 15, 313–389.

LAWSON, J D. 1954. The Silurian succession at Gorsley (Herefordshire). *Geological Magazine,* Vol. 91, 227–237.

LAWSON, J D. 1982. Ludlow stratigraphy at Ludlow, Shropshire. *Geological Magazine,* Vol. 119, 617–618.

LAWSON, J D, and WHITE, D E. 1989. The Ludlow Series in the Ludlow area. 73–90 In *A global standard for the Silurian System.* HOLLAND, C H, and BASSETT, M G (editors). *National Museum of Wales, Geological Series,* No.9.

LEE, M K, PHARAOH, T C, and SOPER, N J. 1990. Structural trends in central Britain from images of gravity and aeromagnetic fields. *Journal of the Geological Society of London,* Vol. 147, 241–258.

LEE, M K, PHARAOH, T C, and GREEN, C A. 1991. Structural trends in the concealed basement of eastern England from images of regional potential field data. *Annales de la Société Géologique de Belgique,* Vol. 114, 45–62.

LE MAITRE, R W. 1976. Some problems of the projection of chemical data into mineralogical classifications. *Contributions to Mineralogy and Petrology,* Vol. 56, 181–189.

LESLIE, A B, SPIRO, B, and TUCKER, M E. 1993. Geochemical and mineralogical variations in the upper Mercia Mudstone Group (Late Triassic), southwest Britain: correlation of outcrop sequences with borehole geophysical logs. *Journal of the Geological Society of London,* Vol. 150, 67–75.

LEWIS, C E, GREEN, P F, CARTER, A, and HURFORD, A J. 1992. Elevated K/T palaeotemperatures throughout northwest England: three kilometres of Tertiary erosion? *Earth and Planetary Science Letters,* Vol. 112, 131–145.

LOTT, G K, SOBEY, R A, WARRINGTON, G, and WHITTAKER, A. 1982. The Mercia Mudstone Group (Triassic) in the Wessex Basin. *Proceedings of the Ussher Society,* Vol. 5, 340–346.

LUCY, W C. 1872. The gravels of the Severn, Avon and Evenlode, and their extension over the Cotteswold Hills. *Proceedings of the Cotteswold Naturalists' Field Club,* Vol. 5, 71–142.

MABILLARD, J–E. 1981. Micropalaeontology and correlation of the Llandovery-Wenlock boundary beds in Wales and the Welsh Borderland. Unpublished PhD thesis, University of Nottingham.

MABILLARD, J–E, and ALDRIDGE, R J. 1985. Microfossil distribution across the base of the Wenlock Series in the type area. *Palaeontology,* Vol. 28, 89–100.

MACKIE, G E. 1887. *Midland Naturalist,* Vol. 10, 197–198.

MADDY, D, KEEN, D H, BRIDGLAND, D R, and GREEN, C P. 1991. A revised model for the Pleistocene development of the River Avon, Warwickshire. *Journal of the Geological Society of London,* Vol. 148, 473–484.

MADDY, D, GREEN, C P, LEWIS, S G, and BOWEN, D Q. 1993. Pleistocene geology of the Lower Severn Valley, UK. *Quaternary Science Reviews*, Vol. 14, 209–222.

MATLEY, C A. 1912. The Upper Keuper (or Arden) Sandstone Group, and associated rocks of Warwickshire. *Quarterly Journal of the Geological Society of London*, Vol. 68, 252–282.

MCKEE, J R. 1973. *Worcester City: an urban structure plan.* (Russell of Worcester.)

MESCHEDE, M. 1986. A method of discriminating between different types of mid-ocean ridge basalts and continental tholeiites with the Nb-Zr-Y diagram. *Chemical Geology*, Vol. 56, 207–218.

MILLSON, J A. 1987. The Jurassic evolution of the Celtic Sea basins. 599–610 in *Petroleum geology of North West Europe*, Volume 2. BROOKS, J, and GLENNIE, K W (editors). (London: Graham & Trotman.)

MINSHULL, G N. 1974. The City of Worcester. 12–17 in *Worcester and its region. Field studies in the former county of Worcestershire*. ADLAM, B H (editor). (Worcester Geographical Association.)

MITCHELL, G H, POCOCK, R W, and TAYLOR, J H. 1962. Geology of the country around Droitwich, Abberley and Kidderminster. *Memoir of the Geological Survey of Great Britain*, Sheet 182 (England and Wales).

MITCHELL, F G, PENNY, L F, SHOTTON, F W, and WEST, R G. 1973. A correlation of Quaternary deposits in the British Isles. *Special Report of the Geological Society of London*, No. 4.

MOORLOCK, B S P, BARRON, A J M, AMBROSE, K, and CANNELL, B. 1985. *Geology of Sheet SO 85.* (Keyworth: British Geological Survey.)

MORRIS, L. 1974. The site of Worcester: its geology and geomorphology. 23–34 in *Worcester and its region. Field studies in the former county of Worcestershire*. ADLAM, B H (editor). (Worcester Geographical Association.)

MORRIS, L. 1984. Stability of detrital heavy minerals in Tertiary sandstones of the North Sea Basin. *Clay Minerals*, Vol. 19, 287–308.

MORRIS, L. 1988. Stratigraphic relationships and provenance of Lower Old Red Sandstone samples from Herefordshire assessed by heavy mineral analysis. *British Geological Survey Technical Report, Stratigraphy Series*, WH/88/7OR.

MUDGE, D C. 1978. Stratigraphy and sedimentation of the Lower Inferior Oolite of the Cotswolds. *Quarterly Journal of the Geological Society of London*, Vol. 135, 611–627.

MURCHISON, R I. 1834. *Outline of the geology of the neighbourhood of Cheltenham.* 40 pp. (Cheltenham: Davies.)

MURCHISON, R I. 1839. *The Silurian System.* (London: J Murray.)

MURCHISON, R I. 1845. *Outline of the geology of the neighbourhood of Cheltenham.* New edition revised and augmented by J Buckman and H E Strickland. (London: John Murray.)

MURCHISON, R I, and STRICKLAND, H E. 1840. On the upper formations of the New Red Sandstone System in Gloucestershire, Worcestershire and Warwickshire, etc. *Transactions of the Geological Society of London*, Series 2, Vol. 5, 331–348.

OLD, R A. 1987. Geological notes and local details for 1:10 000 sheets: SO 94SW (Eckington). (Keyworth: British Geological Survey.)

OLD, R A, HAMBLIN, R J O, AMBROSE, K, and WARRINGTON, G. 1991. Geology of the country around Redditch. *Memoir of the British Geological Survey*, Sheet 183 (England and Wales).

OLD, R A, SUMBLER, M G, and AMBROSE, K. 1987. Geology of the country around Warwick. *Memoir of the British Geological Survey*, Sheet 184 (England and Wales).

OLDHAM, R D. 1894. A comparison of the Permian breccias of the Midlands with the Upper Carboniferous glacial deposits of India and Australia. *Quarterly Journal of the Geological Society of London*, Vol. 50, 463–471.

OWEN, T R. 1954. The structure of the Neath Disturbance between Bryniau Gleison and Glynneath, South Wales. *Quarterly Journal of the Geological Society of London*, Vol. 109 (for 1953), 333–365.

PALMER, R C. 1976. Soils in Herefordshire IV. *Soil Survey, Harpenden, Soil Survey Record*, No. 36.

PARIS, E T. 1908. Notes on some echinoids from the Lias of Worcestershire, Gloucestershire and Somerset. *Proceedings of the Cotteswold Naturalists' Field Club*, Vol. 16, 143–150.

PARKER, A, ALLEN, J R L, and WILLIAMS, B P J. 1983. Clay mineral assemblages of the Townsend Tuff Bed (Lower Old Red Sandstone), South Wales and the Welsh Borders. *Journal of the Geological Society of London*, Vol. 140, 769–779.

PATCHETT, P J, GALE, N H, GOODWIN, R, and HUMM, M J. 1980. Rb-Sr whole-rock isochron ages of late Precambrian to Cambrian igneous rocks from southern Britain. *Journal of the Geological Society of London*, Vol. 137, 649–656.

PEARCE, J A. 1982. Trace element characteristics of lavas from destructive plate boundaries. 525–548 in *Orogenic andesites and related rocks*. THORPE, R S (editor). (Chichester: Wiley and Sons.)

PEARCE, J A, and CANN, J R. 1973. Tectonic setting of basic volcanic rocks determined using trace element analyses. *Earth and Planetary Science Letters*, Vol. 19, 290–300.

PEARCE, J A, HARRIS, N B W, and TINDALE, A G. 1984. Trace element discrimination diagrams for the tectonic interpretation of granitic rocks. *Journal of Petrology*, Vol. 25, 956–983.

PENN, I E. 1987. Geophysical logs in the stratigraphy of Wales and adjacent offshore and onshore areas. *Proceedings of the Geologists' Association*, Vol. 98, 275–314.

PENN, J S W. 1969. The Silurian rocks to the west of the Malvern Hills from Clencher's Mill to Knightsford Bridge. Unpublished PhD thesis, University of London.

PENN, J S W. 1971. Bioherms in the Wenlock Limestone of the Malvern area (Herefordshire, England). *Memoires du Bureau de Recherche Géologique et Miniéres*, No. 73, 129–137.

PENN, J S W, and FRENCH, J. 1971. The Malvern Hills. *Geologists' Association Guide*, No. 4.

PHARAOH, T C, and EVANS, C J (compilers). 1987. Morley Quarry No. 1 geological well completion Report Investigation of the geothermal potential of the UK. (Keyworth, Nottingham: British Geological Survey.)

PHARAOH, T C, LEE, M K, EVANS, C J, and BREWER, T S. 1991. A cryptic late Proterozoic island arc and marginal basin complex in the heart of England. *TERRA Abstracts*, No. 3, 58.

PHAROAH, T C, WEBB, P C, THORPE, R S, and BECKINGSALE, R D. 1987. Geochemical evidence for the tectonic setting of late Proterozoic volcanic suites in central England. 541-542 *In* Geochemistry and mineralization of Proterozoic volcanic suites. PHAROAH, T C, BECKINSALE, R D, and RICKARD, D (editors.) *Special Publication of the Geological Society of London*, No. 33.

PHILLIPS, J. 1848. The Malvern Hills compared with the Palaeozoic districts of Abberley, Woolhope, May Hill, Tortworth and Usk. *Memoir of the Geological Survey*, Vol. 2, Pt 1.

PHIPPS, C B. 1957. The structure and statigraphy of the Silurian rocks west of the south Malvern Hills. Unpublished PhD thesis, University of Birmingham.

PHIPPS, C B. 1962. The revised Ludlovian stratigraphy of the type area — a discussion. *Geological Magazine*, Vol. 99, 385–392.

PHIPPS, C B. 1963. Ludlovian stratigraphy. *Geological Magazine*, Vol. 100, 186.

PHIPPS, C B, and REEVE, F A E. 1964. The Pre-Cambrian–Palaeozoic boundary of the Malverns. *Geological Magazine*, Vol. 101, 397–408.

PHIPPS, C B, and REEVE, F A E. 1967. Stratigraphy and geological history of the Malvern, Abberley and Ledbury Hills. *Geological Journal*, Vol. 5, 339–368.

PHIPPS, C B, and REEVE, F A E. 1969. Structural geology of the Malvern, Abberley and Ledbury hills. *Quarterly Journal of the Geological Society of London*, Vol. 125, 1–37.

PLATT, J I. 1933. The geology of the Warren House Series. *Geological Magazine*, Vol. 70, 423–429.

POCOCK, R W. 1930. The *Petalocrinus* limestone horizon at Woolhope, Herefordshire. *Quarterly Journal of the Geological Society of London*, Vol. 86, 50–63.

POLLARD, J E. 1985. *Isopodichnus*, related arthropod trace fossils and notostracans from Triassic fluvial sediments. *Transactions of the Royal Society of Edinburgh*, Vol. 76, 273–285.

POOLE, E G. 1969. The Putcheons Farm (1965) Borehole, Redditch, Worcestershire. *Bulletin of the Geological Survey of Great Britain*, No. 29, 105–114.

POOLE, E G, and WILLIAMS, B J. 1981. The Keuper saliferous beds of the Droitwich area. *Report of the Institute of Geological Sciences*, No. 81/2.

RAMSBOTTOM, W H C, CALVER, M A, EAGAR, R M C, HODSON, F, HOLLIDAY, D W, STUBBLEFIELD, C J, and WILSON, R B. 1978. A correlation of Silesian rocks in the British Isles. *Special Report of the Geological Society of London*, No. 10.

RAMSEY, A C. 1855. On the occurrence of angular, subangular, polished and striated fragments and boulders in the Permian Breccia of Shropshire, Worcestershire; and on the probable existence of glaciers and icebergs in the Permian epoch. *Quarterly Journal of the Geological Society of London*, Vol. 11, 185–205.

RAW, F. 1952. Structure and origin of the Malvern Hills. *Proceedings of the Geologists' Association*, Vol. 63, 227–239.

READING, H G, and POOLE, A B. 1961. A Llandovery shoreline from the southern Malverns. *Geological Magazine*, Vol. 98, 295–300.

READING, H G, and POOLE, A B. 1962. Malvern structures. *Geological Magazine*, Vol. 99, 377–399.

REEVE, F A E. 1953. The structure of the Silurian rocks of the Malvern and Abberley Hills, Worcestershire. Unpublished PhD thesis, University of Birmingham.

RICHARDSON, J B, and RASUL, S M. 1990. Palynofacies in a Late Silurian regressive sequence in the Welsh Borderland and Wales. *Journal of the Geological Society of London*, Vol. 147, 675–686.

RICHARDSON, J B, and RASUL, S M. 1991. Reply in discussion on palynofacies in a Late Silurian regressive sequence in the Welsh Borderland and Wales. *Journal of the Geological Society of London*, Vol. 148, 783–784.

RICHARDSON, L. 1902. On the sequence of the Inferior Oolite Deposits at Bredon Hill, Worcestershire. *Geological Magazine*, Vol. 9, 513–514.

RICHARDSON, L. 1903a. The Rhaetic rocks of north-west Gloucestershire. *Proceedings of the Cotteswold Naturalists' Field Club*, Vol. 14, 127–174.

RICHARDSON, L. 1903b. On two sections of the Rhaetic rocks in Worcestershire. *Geological Magazine*, Vol. 40, 80–82.

RICHARDSON, L. 1904a. The Rhaetic rocks of Worcestershire. *Proceedings of the Cotteswold Naturalists' Field Club*, Vol. 15, 19–44.

RICHARDSON, L. 1904b. The evidence for a non-sequence between the Keuper and Rhaetic Series in north-west Gloucestershire and Worcestershire. *Quarterly Journal of the Geological Society of London*, Vol. 60, 349–358.

RICHARDSON, L. 1905. Notes on the geology of Bredon Hill. *Transactions of the Woolhope Naturalists' Field Club*, Vol. for 1902, 62–67.

RICHARDSON, L. 1906. The Lias of Worcestershire. *Transactions of the Worcester Naturalists' Field Club*, Vol. 3, 188–206.

RICHARDSON, L. 1907. An outline of the geology of Herefordshire. *Transactions of the Woolhope Naturalists' Field Club*, Vol. for 1905, 1–68.

RICHARDSON, L. 1923. A boring in the Trias at Upton-on-Severn, Worcs. *Geological Magazine*, Vol. 60, 119–121.

RICHARDSON, L. 1924. Bredon Hill: its geological structure and water resources. *Transactions of the Worcester Naturalists' Field Club*, Vol. 8, 82–88.

RICHARDSON, L. 1929. The country around Moreton in Marsh. *Memoir of the Geological Survey of Great Britain*, Sheet 217, (England and Wales.)

RICHARDSON, L. 1930. Wells and springs of Worcestershire. *Memoir of the Geological Survey of Great Britain*.

RICHARDSON, L. 1935. Wells and springs of Herefordshire. *Memoir of the Geological Survey of Great Britain*.

RICHARDSON, L. 1948. The upper limit of the Rhaetic Series and the relationship of the Rhaetic and Liassic Series. *Proceedings of the Cotteswold Naturalists' Field Club*, Vol. 29, 143–144.

RICHARDSON, L. 1956. The geology of Worcester. *Transactions of the Worcestershire Naturalists' Club*, Vol. 11, 29–65.

RICHARDSON, L. 1964. The upper limit of the Rhaetic Series and the relationship of the Rhaetic and Liassic Series: a correction. *Proceedings of the Cotteswold Naturalists' Field Club*, Vol. 34, 153.

ROBERTSON, T. 1926. The section of the new railway tunnel through the Malvern Hills at Colwall. *Summary of Progress of the Geological Survey of Great Britain and the Museum of Practical Geology for the Year 1925*, 162–173.

ROSE, J. 1987. Status of the Wolstonian glaciation in the British Quaternary. *Quaternary Newsletter*, Vol. 53, 1–9.

SAUNDERS, A D, and TARNEY, J. 1984. Geochemical characteristics of basaltic volcanism within back-arc basins. 59–76 in *Marginal basin geology*. KOKELAAR, B P, and HOWELLS, M F (editors). *Special Publication of the Geological Society of London*, No. 16.

SCOFFIN, T P. 1971. The conditions of growth of the Wenlock reefs of Shropshire (England). *Sedimentology*, Vol. 17, 173–219.

Sedgwick, A. 1853. On a proposed separation of the so-called Caradoc Sandstone into two distinct groups; viz (1) May Hill Sandstone; (2) Caradoc Sandstone. *Quarterly Journal of the Geological Society of London*, Vol. 9, 215–230.

Shelford, P H. 1964. The Malvern line. *Geological Magazine*, Vol. 101, 566–567.

Sherlock, R L. 1926. A correlation of the British Permo-Triassic rocks — I. North England, Scotland and Ireland. *Proceedings of the Geologists' Association*, Vol. 37, 1–69.

Shotton, F W. 1937. The Lower Bunter Sandstones of North Worcestershire and East Shropshire. *Geological Magazine*, Vol. 74, 534–553.

Shotton, F W. 1968. The Pleistocene succession around Brandon, Warwickshire. *Philosophical Transactions of the Royal Society of London*, Series B, Vol. 254, 387–400.

Shotton, F W. 1973. The English Midlands. 18–22 *in* A correlation of the Quaternary deposits in the British Isles. Mitchell, G F, Penny, L F, Shotton, F W, and West, R G (editors). *Special Report of the Geological Society of London*, No.4.

Shotton, F W. 1983. Interglacials after the Hoxnian in Britain. 109–115 in *Geological Society of London Project 73/1/24. Quaternary Glaciations in the Northern Hemisphere Report 10.* Billard, A, Conchon, O, and Shotton, F W (editors). (Paris: UNESCO.)

Shotton, F W. Banham, P H, and Bishop, W W. 1977. Glacial–interglacial stratigraphy of the Quaternary in Midland and Eastern England. 267–282 in *British Quaternary studies: recent advances*. Shotton, F W (editor). (Oxford: Clarendon Press.)

Simms, M J. 1990. Upper Pleinsbachian stratigraphy in the Severn Basin area: evidence for anomalous structural controls in the Lower and Middle Jurassic. *Proceedings of the Geologists' Association*, Vol. 101, 131–144.

Simms, M J. and Ruffell, A H. 1990. Climate and biotic change in the late Triassic. *Journal of the Geological Society of London*. Vol. 147, 321–327.

Siveter, D J, Owens, R M, and Thomas, A T. 1989. Silurian field excursions: a geotraverse across Wales and the Welsh Borderland. *National Museum of Wales, Geological Series*, No. 10.

Smith, D B, Brunstrom, R G W, Manning, P I Simpson, S, and Shotton, F W. 1974. A correlation of Permian rocks in the British Isles. *Special Report of the Geological Society of London*, No. 5.

Smith, I F. 1986. Mesozoic basins. 42–83 in *Geothermal energy — the potential in the United Kingdom*. Downing, R A, and Gray, D A (editors). (London: HMSO for British Geological Survey.)

Smith, I F. and Burgess, W G. 1984. The Permo-Triassic rocks of the Worcester Basin: Investigation of the geothermal potential of the UK. (Keyworth, Nottingham: British Geological Survey.)

Smith, N J P. 1985 (compiler). *Pre-Permian geology of the United Kingdom (south)*. Scale 1:1 000 000. Two maps commemorating the 150th Anniversary of the British Geological Survey. (Mitcham, Surrey: Cook, Hammond & Kell for the British Geological Survey.)

Smith, R D A, and Ainsworth, R B. 1989. Hummocky cross-stratification in the Downton of the Welsh Borderland. *Journal of the Geological Society of London*, Vol. 146, 897–900.

Soper, N J, and Woodcock, N H. 1990. Silurian collision and sediment dispersal patterns in southern Britain. *Geological Magazine*, Vol. 127, 527–542.

Spinner, E. 1966. Palynological evidence on the age of the Carboniferous beds of Woodbury Hill, near Abberley, Worcestershire. *Proceedings of the Yorkshire Geological Society*, Vol. 35, 507–522.

Squirrell, H C, and Tucker, E V. 1960. The geology of the Woolhope inlier (Herefordshire). *Quarterly Journal of the Geological Society of London*, Vol. 116, 139–185.

Stamp, L D. 1923. The base of the Devonian with special reference to the Welsh Borderland. *Geological Magazine*, Vol. 60, 367–372.

Stensiö, E A. 1932. *The Cephalaspids of Britain.* (London: British Museum (Natural History).)

Strachan, R A, Nance, R D, Dallmeyer, R D, D'Lemos, R S, Murphy, J B, and Watt, G R. 1996. Late Precambrian tectonothermal evolution of the Malverns Complex. Journal of the Geological Society, London, Vol. 153, 589–600.

Strange, P J, and Ambrose, K. 1982. *Geological notes and local details for 1:10 000 sheets: SP 16NW, NE, SW, SE and parts of SP 15NW (Henley-in-Arden.)* (Keyworth: Institute of Geological Sciences.)

Streckeisen, A. 1967. Classification and nomenclature of igneous rocks. *Neues Jahrbuch für Mineralogie*, Vol. 107, 144–214.

Strickland, A E. 1842. Memoir descriptive of a series of coloured sections of the cuttings on the Birmingham and Gloucester railway. *Transactions of the Geological Society of London*, Series 2, Vol. 6, 545–555.

Strong, G E. 1986. Petrographical notes on thin sections from localities in Herefordshire. *British Geological Survey Technical Report*, WH/88/84R.

Stuart, A J. 1982. *Pleistocene vertebrates in the British Isles.* (London: Longman.)

Sumbler, M G. 1983a. A new look at the type Wolstonian glacial deposits of Central England. *Proceedings of the Geologists' Association*, Vol. 94, 21–31.

Sumbler, M G. 1983b. The type Wolstonian sequence — some further comments. *Quaternary Newsletter*, Vol. 40, 36–39.

Symonds, W S. 1872. *Records of the rocks.* (London: John Murray.)

Symonds, W S. 1883. *The Severn Straits: notes on glacial drifts, bone caverns, and old glaciers, some within reach of the Malvern Hills.* (Tewkesbury: William North; London: Simpkin, Marshall & Co.)

Symonds, W S. 1907. The occurrence of oolitic remains in the gravel of Cradley. *Proceedings of the Cotteswolds Naturalists' Field Club*, Vol. for 1853, 48–49.

Symonds, W S, and Lambert, A. 1861. Sections of Malvern and Ledbury tunnels and the intervening line of railroad. *Quarterly Journal of the Geological Society of London*, Vol. 17, 152–160.

Thorpe, R S. 1971. A potash-rich trachyte from the Precambrian of England. *Contributions to Mineralogy and Petrology*, Vol. 31, 115–120.

Thorpe, R S. 1972. Possible subduction zone origins for two Precambrian calc-alkaline complexes from southern Britain. *Bulletin of the Geological Society of America*, Vol. 83, 3663–3668.

Thorpe, R S. 1974. Aspects of magmatism and plate tectonics in the Precambrian of England and Wales. *Geological Journal*, Vol. 9, 115–135.

THORPE, R S. 1987. Pseudotachylite from a Precambrian shear zone in the Malvern Hills. *Proceedings of the Geologists' Association*, Vol. 98, 205–210.

THORPE, R S, BECKINSALE, R D, PATCHETT, P J, PIPER, J D A, DAVIES, G R, and EVANS, J A. 1984. Crustal growth and late Precambrian–early Palaeozoic plate tectonic evolution of England and Wales. *Journal of the Geological Society of London*, Vol. 141, 521–536.

TOMLINSON, M E. 1925. River terraces of the lower valley of the Warwickshire Avon. *Quarterly Journal of the Geological Society of London*, Vol. 81, 137–163.

TOMLINSON, M E. 1941. Pleistocene gravels of the Cotswold sub-edge plain from Mickleton to the Frome Valley. *Quarterly Journal of the Geological Society of London*, Vol. 96, 385–420.

TOWNROW, J A. 1962. On some disaccate pollen grains of Permian to middle Jurassic age. *Grana Palynologica*, Vol. 3, 13–44.

TRUSHEIM, F. 1963. Zur Gliederung des Buntsandsteins. *Erdöl-Zeitschrift für Bohr- und Fordertechnik*, Vol. 79, 277–292.

TUCKER, E V. 1965. The Malvern line. *Geological Magazine*, Vol. 102, 88–90.

TUCKER, R D, and PHARAOH, T C. 1991. U-Pb zircon ages for Late Precambrian igneous rocks in southern Britain. *Journal of the Geological Society of London*, Vol. 148, 435–443.

TUNBRIDGE, I P. 1983. Geophysical down-hole recognition of the Lower Devonian 'Psammosteus' Limestone and Townsend Tuff Bed, South Wales. *Geological Journal*, Vol. 18, 325–329.

TUTTLE, O F, and BOWEN, N L. 1958. Origin of granite in the light of experimental studies in the system $NaAlSi_3O_8$–$KAlSi_3O_8$–SiO–H_2O. *Memoir of the Geological Society of America*, 74.

VISSCHER, H, and BRUGMAN, W A. 1981. Ranges of selected palynomorphs in the Alpine Triassic of Europe. *Review of Palaeobotany and Palynology*, Vol. 34, 115–128.

WARRINGTON, G. 1967. Correlation of the Keuper Series of the Triassic by miospores. *Nature, London*, Vol. 214, 1323–1324.

WARRINGTON, G. 1968. The stratigraphy and palaeontology of the 'Keuper' Series in Central England (Worcestershire, Warwickshire, Staffordshire and Leicestershire). Unpublished PhD thesis, University of London.

WARRINGTON, G. 1970a. The stratigraphy and palaeontology of the 'Keuper' Series of the central Midlands of England. *Quarterly Journal of the Geological Society of London*, Vol. 126, 183–223.

WARRINGTON, G. 1970b. The "Keuper Series" of the British Trias in the northern Irish Sea and neighbouring areas. *Nature, London*, Vol. 226, 254–256.

WARRINGTON, G. 1974. Studies in the palynological biostratigraphy of the British Trias. I. Reference sections in west Lancashire and north Somerset. *Review of Palaeobotany and Palynology*, Vol. 17, 133–147.

WARRINGTON, G. 1978. Palynology of the Keuper, Westbury and Cotham beds and the White Lias of the Withycombe Farm Borehole. *Bulletin of the Geological Survey of Great Britain*, No. 68, 22–28.

WARRINGTON, G, AUDLEY-CHARLES, M G, ELLIOTT, R E, EVANS, W B, IVIMEY-COOK, H C, KENT, P E, ROBINSON, P L, SHOTTON, F W, and TAYLOR, F W. 1980. A correlation of Triassic rocks in the British Isles. *Special Report of the Geological Society of London*, No. 13.

WARRINGTON, G, and IVIMEY-COOK, H C. 1992. Triassic. 97–106 *in* Atlas of palaeogeography and lithofacies. COPE, J C W, INGHAM, J K, and RAWSON, P F. (editors). *Memoir of the Geological Society of London*, No. 13.

WATKINS, R. 1979. Benthic community organization in the Ludlow Series of the Welsh Borderland. *Bulletin of the British Museum of Natural History (Geology)*, Vol. 31, 175–280.

WATKINS, R, and AITHIE, C J. 1980. Carbonate shelf environments and faunal communities in the Upper Bringewood Beds of the British Silurian. *Palaeogeography, Palaeoclimatology, Palaeoecology*, Vol. 29, 341–368.

WHITE, D E, ELLISON, R A, and MOORLOCK, B S P. 1984. New information on the stratigraphy of the upper Silurian rocks in the southern part of the Malverns district, Hereford and Worcester. *BGS Short Communications 5. Report of the British Geological Survey*, Vol. 16, No. 1, 13–18.

WHITE, D E, and LAWSON, J D. 1989. The Přídolí Series in the Welsh Borderland and south-central Wales. 131-141 *in* A global standard for the Silurian System. HOLLAND, C H, and BASSETT, M G (editors). *National Museum of Cardiff, Wales, Geological Series*, No. 9.

WHITE, E I. 1950. A fish from the Bunter near Kidderminster. *Transactions of the Worcestershire Naturalists' Club*, Vol. 10, 185–189.

WHITEHEAD, P F. 1989a. The development of the Avon Valley river terraces. 37–41 in *The Pleistocene of the West Midlands: field guide*. KEEN, D H (editor). (Cambridge: Quaternary Research Association.)

WHITEHEAD, P F. 1989b. The Quaternary malacofauna of the Warwickshire–Worcestershire Avon. 42–48 in The Pleistocene of the West Midlands: field guide. KEEN, D H (editor). (Cambridge: Quaternary Research Association.)

WHITTAKER, A. 1972a. Intra-Liassic structures in the Severn Basin area. *Report of the Institute of Geological Sciences*, No. 72/3.

WHITTAKER, A. 1972b. Geology of Bredon Hill, Worcestershire. *Bulletin of the Geological Survey of Great Britain*, Vol. 42, 1–49.

WHITTAKER, A. 1980. Kempsey No. 1 Geological Well Completion Report. *Report of the Deep Geology Unit, Institute of Geological Sciences*, No. 80/1.

WHITTAKER, A. (editor). 1985. *Atlas of onshore sedimentary basins in England and Wales: post-Carboniferous tectonics and stratigraphy*. (Glasgow and London: Blackie.)

WHITTAKER, A, and GREEN, G W. 1983. Geology of the country around Weston-super-Mare. *Memoir of the Geological Survey of Great Britain*, Sheet 279 with parts of sheets 263 and 295 (England and Wales.)

WHITTAKER, A, HOLLIDAY, D W, and PENN, I E. 1985. Geophysical logs in British stratigraphy. *Special Report of the Geological Society of London*, No. 18. 74 pp.

WHITWORTH, T. 1962. Malvern structures. *Geological Magazine*, Vol. 99, 375–377.

WICKHAM, W. 1899. On the drifts of our local rivers and brooks, with a short sketch of the sands at Mathon. *Transactions of the Malvern Field Club*, 5, Vol. for 1899, 3–15.

WILLE, W. 1970. *Plaesiodictyon mosellanum* n.g., n.sp., eine mehrzellige Grünalge aus dem Unteren Keuper von Luxemburg. *Neues Jahrbuch für Geologie und Paläontologie Monatshefte*, Jg 1970, H.5, 283–310.

WILLIAMS, B J, and WHITTAKER, A. 1974. Geology of the country around Stratford-upon-Avon and Evesham. *Memoir of the Geological Survey of Great Britain*, Sheet 200 (England and Wales).

WILLS, L J. 1910. On the fossiliferous lower Keuper rocks of Worcestershire, with descriptions of some plants and animals discovered therein. *Proceedings of the Geologists' Association,* Vol. 21, 249–331.

WILLS, L J. 1924. The development of the Severn valley in the neighbourhood of Ironbridge and Bridgnorth. *Quarterly Journal of the Geological Society of London,* Vol. 80, 274–314.

WILLS, L J. 1938. The Pleistocene development of the Severn from Bridgnorth to the sea. *Quarterly Journal of the Geological Society of London,* Vol. 94, 161–242.

WILLS, L J. 1948. *The palaeogeography of the Midlands.* (University Press of Liverpool.)

WILLS, L J. 1956. *Concealed coalfields.* (Glasgow and London: Blackie & Son.)

WILLS, L J. 1970. The Triassic succession in the English Midlands in its regional setting. *Quarterly Journal of the Geological Society of London,* Vol. 26, 225–285.

WILLS, L J. 1976. The Trias of Worcestershire and Warwickshire. *Report of the Institute of Geological Sciences,* No. 76/2.

WILLS, L J, and CAMPBELL-SMITH, W. 1913. Notes on the flora and fauna of the Upper Keuper Sandstones of Warwickshire and Worcestershire. *Geological Magazine,* Vol. 50, 461–462.

WILSON, A A. 1990. The Mercia Mudstone Group (Trias) of the East Irish Sea Basin. *Proceedings of the Yorkshire Geological Society,* Vol. 48, 1–22.

WILSON, A A, and EVANS, W B. 1990. Geology of the country around Blackpool. *Memoir of the British Geological Survey,* Sheet 66 (England and Wales).

WOODCOCK, N H. 1984. Early Palaeozoic sedimentation and tectonics in Wales. *Proceedings of the Geologists' Association,* Vol. 95, 323–335.

WOODCOCK, N H. 1990. Sequence stratigraphy of the Palaeozoic Welsh Basin. *Journal of the Geological Society,* Vol. 147, 537–547.

WOODWARD, A S. 1891. Catalogue of the fossil fishes in the British Museum (Natural History), Vol. 2.

WOODWARD, H B. 1893. The Jurassic rocks of Britain: Vol. III. The Lias of England and Wales (Yorkshire excepted). *Memoir of the Geological Survey of Great Britain.*

WOODWARD, H B. 1894. The Jurassic rocks of Britain: Vol. IV. The Lower Oolitic Rocks of England and Wales (Yorkshire excepted). *Memoir of the Geological Survey of Great Britain.*

WORSSAM, B C, ELLISON, R A, and MOORLOCK, B S P. 1989. Geology of the country around Tewksbury. *Memoir of the British Geological Survey,* Sheet 216 (England and Wales).

WRIGHT, T. 1860. On the zone of *Avicula contorta* and the Lower Lias of the south of England. *Quarterly Journal of the Geological Society of London,* Vol. 16, 374–411.

WRIGHT, T. 1878–1886. Monograph of the Lias ammonites of the British Islands. *Monograph of the Palaeontographical Society of London,* Parts 1–8, pp.1–503. (1878, pp.1–48, Pls i–viii; 1882, pp.329–400, pls xlix–lxix.)

WRIGHT, V P, NORTH, C P, HANCOCK, P L, CURTIS, M, and ROBINSON, D. 1988. Pedofacies variations across an arid alluvial basin: a case study from the Upper Triassic of SW Britain [Abstract]. 227–228 in *International Association of Sedimentologists European Meeting 1988, Leuven.*

ZIEGLER, A M. 1964. The Malvern Line. *Geological Magazine,* Vol. 101, 467–469.

ZIEGLER, A M. 1965. Silurian marine communities and their environmental significance. *Nature, London,* Vol. 207, 270–272.

ZIEGLER, A M, COCKS, L R M, and BAMBACH, R K. 1968. The composition and structure of Lower Silurian marine communities. *Lethaia,* Vol. 1, 1–27.

ZIEGLER, A M, COCKS, L R M, and McKERROW, W S. 1968. The Llandovery transgression of the Welsh Borderland. *Palaeontology,* Vol. 11, 736–782.

APPENDIX 1

Open-file technical reports

The following lists the open-file technical reports which give details of the geology within each 1:10 000 map area. They can be consulted at BGS libraries or purchased from the Sales Desk, BGS, Keyworth, Nottingham. Those marked * are part of an earlier series termed 'Geological notes and local details for 1:10 000 sheets'. Authors are named in the list of 1:10 000 maps (p.xii).

SO 63 NE	Little Marcle	BSPM	WA/92/74
SO 64 NE	Bishop's Frome	WJB	WA/90/57
SO 64 SE	Bosbury	AB	*
SO 65 NE	Tedstone Delamere	WJB, AAJ	*
SO 65 SE	Stanford Bishop	PJS, PAR	WA/90/84
SO 73 NW	Ledbury	BSPM	WA/90/85
SO 73 NE	South Malvern Hills	BSPM	WA/90/86
SO 74 NW	Cradley	WJB	WA/90/89
SO 74 NE	Great Malvern	WJB	WA/90/87
SO 74 SW	Coddington	AB	WA/89/3
SO 74 SE	Malvern Wells	BSPM	WA/90/88
SO 75 NW	Whitbourne	WJB, PAR	WA/90/90
SO 75 NE	Kenswick	PAR	*
SO 75 SW	Suckley	PAR	WA/90/91
SO 75 SE	Leigh Sinton	PAR	*
SO 83 NW	Longdon	BSPM	*
SO 83 NE	Ripple	BSPM, RJW	*
SO 84 NW	Madresfield	BSPM	WA/92/87
SO 84 NE	Kempsey	KA	WA/88/25
SO 84 SW	Hanley Swan	BSPM	WA/92/76
SO 84 SE	Earls Croome	KA	WA/88/24
SO 85 NW	Worcester NW	KA, AJMB, AB	WA/88/26

SO 85 NE	Worcester NE	BSPM	WA/92/77
SO 85 SW	Worcester SW	AJMB	*
SO 85 SE	Worcester SE	AJMB	*
SO 93 NW	Bredon & Ashchurch	RJW	*
SO 93 NE	Beckford	BCW	*
SO 94 NW	Pershore	RAO	*
SO 94 SW	Eckington	RAO	*
SO 95 NW SO 95 NE (part)	Crowle	KA	*
SO 95 SW SO 95 SE (part)	White Ladies Aston	KA	*

The following reports describe the geology with special emphasis on potential resources of sand and gravel. They are part of a series named 'Geological reports for DoE: Land use planning.

Authors not previously listed are B Cannell (BC), P J Strange (PJS) and M G Sumbler (MGS).

Quaternary deposits of sheet SO 64	PJS, AB
Geology of sheet SO 65	MGS
Geology of sheet SO 84	KA, BSPM, BC
Geology of sheet SO 85	BSPM, AJMB, KA, BC

Other open-file technical report

Details of the Pleistocene deposits west of the Malvern Hills. Report WA/90/92. W J Barclay, A Brandon, R A Ellison and B S P Moorlock.

APPENDIX 2

Geochemical analysis of the Precambrian rocks

The samples were cleaned, jaw-crushed and ground to a fine powder in an agate mill. All samples were analysed by X-ray fluorescence (XRF) spectrometry using the Philips PW1400 instruments at Nottingham University and Midland Earth Science Associates. Major elements (SiO_2, TiO_2, Al_2O_3, Fe_2O_3, MnO, MgO, CaO, Na_2O, K_2O, P_2O_5) were determined on fused glass beads. The values are expressed in weight %. Where a value for FeO is shown, it has been calculated by Dearnley (1990) using the oxidation ratio adjustment formula recommended by Le Maitre (1976). LOI, loss on ignition at 1050°C; Rest, trace elements expressed in weight %. Trace elements (Ba, Ce, Co, Cr, Cu, Nb, Ni, Pb, Rb, Sr, Th, V, Y, Zn, Zr) were determined on pressed powder pellets with a PVP/methyl cellulose binder.

1. Major and trace element data for the Malverns Complex (Dearnley, 1990; Pharaoh and Brewer unpublished data)

CUMULATE AMPHIBOLITES (ULTRABASIC–BASIC)

Sample number	58233	58229	58232	58234	58247	M9
Locality	N Malvern Quarries	N Malvern Quarries	N Malvern Quarry	Hollybush* Quarry	Hollybush* Quarry	Tank Quarry
Lithology	Ultrabasic +felsic vein	Gabbro	Gabbro +felsic vein	Gabbro	Gabbro	Gabbro (altered)
Grid reference	SO 7702 4689	SO 7702 4689	SO 7702 4689	SO 7593 3718	SO 7593 3718	SO 7682 4706
SiO_2	43.85	48.29	47.98	47.97	48.42	56.65
TiO_2	1.49	.68	1.50	.70	.48	.68
Al_2O_3	8.77	7.61	13.08	13.15	13.47	12.08
Fe_2O_3	3.20	2.18	3.04	2.21	1.18	8.44
FeO	10.48	6.61	6.50	6.12	6.03	—
MnO	.42	.19	.29	.18	.17	.40
MgO	12.98	15.69	9.77	12.06	11.45	9.48
CaO	12.47	12.77	9.54	11.87	13.24	5.36
Na_2O	.94	1.19	2.70	1.37	.97	2.24
K_2O	.69	.68	1.82	1.18	.28	2.03
P_2O_5	1.49	.30	.82	.05	.08	.11
LOI	2.14	1.44	2.07	2.33	2.97	2.38
Rest	.31	.32	.33	.23	.21	.28
Total	99.23	97.95	99.44	99.42	98.95	100.13
Trace elements in parts per million						
Ba	176	136	784	180	32	431
Ce	231	69	69	18	14	37
Co	50	56	38	47	47	38
Cr	688	1232	479	564	598	738
Cu	48	15	14	70	61	7
Nb	21	12	13	2	—	12
Ni	126	147	98	173	167	217
Pb	6	4	6	7	10	9
Rb	12	15	53	38	7	59
Sr	181	194	362	182	261	171
Th	3	—	1	—	2	7
V	301	201	235	177	135	138
Y	113	29	36	24	25	59
Zn	218	89	161	73	76	136
Zr	89	96	233	109	88	59
Mg′	63.39	76.54	65.34	72.61	74.21	68.99
Nb/Y	.19	.41	.36	.08		.20
Zr/Y	.79	3.31	6.47	4.54	3.52	1.00
Zr/Nb	4.24	8.00	17.92	54.50		4.92
Zr/TiO_2*10 000	.01	.01	.02	.02	.02	.01
Ce/Y	2.04	2.38	1.92	.75	.56	.63
(Ce/Y)n	5.11	5.95	4.79	1.88	1.40	1.57
Th/Nb	.14		.08			.58
K/Rb	477.31	376.31	285.05	257.77	332.04	285.61
K/Ba	32.54	41.51	19.27	54.42	72.63	39.10
Ba/Rb	14.67	9.07	14.79	4.74	4.57	7.31

* Tewkesbury (216) Sheet

MAFIC AMPHIBOLITES (GABBRO–MAFIC DIORITE)

Sample number	58238	58241	58248	58237	58244	58245	58235	58242	58231	M6
Locality	Upper County Quarry	Upper County Quarry	Upper County Quarry	Gullet* Quarry	Hollybush* Quarry	Dingle Quarry	Dingle Quarry	Dingle Quarry	Earnshaw Quarry	Dingle Quarry
Lithology	Gabbro	Gabbro	Gabbro	Gabbro	Gabbro	Gabbro	Gabbro	Gabbro	Gabbro	Diorite
Grid reference	SO 7680 4480	SO 7680 4480	SO 7673 4477	SO 7625 3808	SO 7593 3718	SO 7654 4567	SO 7654 4567	SO 7697 4447	SO 7697 4447	SO 7652 4567
SiO_2	48.85	50.14	50.77	44.02	52.19	49.41	50.38	42.96	42.74	50.42
TiO_2	1.33	1.40	1.35	1.04	.66	.69	.65	.70	.63	.63
Al_2O_3	15.31	14.80	15.43	17.08	14.61	14.98	15.29	16.94	17.28	15.21
Fe_2O_3	3.50	3.17	3.15	3.36	2.32	2.55	2.65	2.47	2.46	8.84
FeO	7.49	7.21	7.35	7.26	5.23	6.10	5.92	7.54	7.49	—
MnO	.28	.22	.22	.13	.19	.17	.15	.13	.13	.16
MgO	5.99	5.81	5.47	10.47	8.28	9.32	9.35	11.60	11.44	9.16
CaO	8.61	9.16	9.88	4.84	8.78	9.35	8.79	11.18	11.32	10.08
Na_2O	2.98	2.53	2.42	.86	2.58	1.33	1.47	.94	.90	1.43
K_2O	1.47	1.36	1.24	3.79	1.27	2.23	2.54	1.20	1.27	1.81
P_2O_5	.14	.13	.12	.09	.10	.16	.18	.05	.05	.15
LOI	1.52	1.88	2.07	4.62	3.42	2.30	2.18	3.44	3.23	1.68
Total	97.67	97.99	99.66	97.76	99.80	98.82	99.77	99.36	99.14	99.80
Trace elements in parts per million										
Ba	394	307	259	268	179	418	453	206	218	591
Ce	36	26	19	11	24	32	20	18	12	24
Co	34	31	35	40	34	37	35	54	57	49
Cr	139	106	113	268	273	300	290	238	230	264
Cu	8	9	12	45	49	91	51	121	105	59
Nb	7	5	5	3	6	5	6	4	3	4
Ni	48	40	39	100	96	97	93	101	95	91
Pb	10	8	10	9	9	8	7	5	7	5
Rb	48	44	41	200	47	103	116	54	57	82
Sr	272	258	307	96	223	305	292	372	382	340
Th	—	—	—	—	3	—	—	—	—	—
V	304	307	316	281	163	197	193	255	244	161
Y	39	35	37	26	27	18	16	12	12	15
Zn	109	103	105	73	79	76	76	64	62	67
Zr	89	91	94	62	82	72	74	55	50	62
Mg′	50.08	50.71	48.90	64.47	66.85	66.43	66.74	67.92	67.75	67.24
Nb/Y	.18	.14	.14	.12	.22	.28	.38	.33	.25	.27
Zr/Y	2.28	2.60	2.54	2.38	3.04	4.00	4.63	4.58	4.17	4.13
Zr/Nb	12.71	18.20	18.80	20.67	13.67	14.40	12.33	13.75	16.67	15.50
Zr/TiO_2*10 000	.01	.01	.01	.01	.01	.01	.01	.01	.01	.01
Ce/Y	.92	.74	.51	.42	.89	1.78	1.25	1.50	1.00	1.60
(Ce/Y)n	2.31	1.86	1.28	1.06	2.22	4.44	3.13	3.75	2.50	4.00
Th/Nb	—	—	—	—	.50	—	—	—	—	—
K/Rb	254.22	256.58	251.05	157.30	224.30	179.72	181.76	184.47	184.95	183.23
K/Ba	30.97	36.77	39.74	117.39	58.90	44.29	46.54	48.36	48.36	25.42
Ba/Rb	8.21	6.98	6.32	1.34	3.81	4.06	3.91	3.81	3.82	7.21

* Tewkesbury (216) Sheet

INTERMEDIATE AMPHIBOLITES (DIORITE−TONALITE)

Sample number	58259	58261	58256	58252	58253	58246
Locality	N Malvern Quarry	N Malvern Quarry	N Malvern Quarry	N Malvern Quarry	N Malvern Quarry	Upper County Quarry
Lithology	Diorite (altered)	Diorite (altered)	Diorite	Diorite	Quartz diorite	Diorite
Grid reference	SO 7702 4689	SO 7702 4689	SO 7715 4685	SO 7715 4685	SO 7715 4685	SO 7680 4480
SiO_2	51.09	51.09	54.64	55.41	57.27	55.67
TiO_2	1.29	1.30	1.12	1.13	.89	1.43
Al_2O_3	20.76	21.25	19.19	19.29	18.59	15.03
Fe_2O_3	2.80	2.69	2.55	2.49	2.27	2.99
FeO	4.02	3.98	3.62	3.65	3.61	6.23
MnO	.10	.09	.06	.07	.09	.25
MgO	3.26	3.23	3.40	3.34	3.15	4.59
CaO	4.30	4.06	4.55	4.59	5.12	6.86
Na_2O	5.38	4.96	4.92	4.25	4.26	2.96
K_2O	2.34	2.47	2.71	3.03	2.19	1.31
P_2O_5	.47	.47	.46	.44	.37	.21
LOI	2.81	3.08	2.36	1.72	2.12	1.83
Rest						
Total	99.01	99.07	99.88	99.70	100.23	99.59
Trace elements in parts per million						
Ba	1134	1213	655	688	741	727
Ce	62	62	44	36	35	37
Co	9	8	19	19	19	25
Cr	7	9	26	23	23	79
Cu	15	14	62	24	13	28
Nb	16	16	6	6	5	15
Ni	4	6	13	7	11	33
Pb	12	9	10	8	11	9
Rb	79	83	78	54	60	45
Sr	848	802	1145	1168	1187	279
Th	—	—	1	1	—	—
V	46	54	123	135	122	241
Y	36	34	13	12	13	42
Zn	121	135	97	87	86	84
Zr	819	840	148	152	140	140
Mg'	47.04	47.35	50.60	50.26	49.83	47.83
Nb/Y	.44	.47	.46	.50	.38	.36
Zr/Y	22.75	24.71	11.38	12.67	10.77	3.33
Zr/Nb	51.19	52.50	24.67	25.33	28.00	9.33
Zr/TiO_2*10 000	.06	.06	.01	.01	.02	.01
Ce/Y	1.72	1.82	3.38	3.00	2.69	.88
(Ce/Y)n	4.31	4.56	8.46	7.50	6.73	2.20
Th/Nb			.17	.17		
K/Rb	245.88	247.03	288.41	465.78	302.99	241.65
K/Ba	17.13	16.90	34.34	36.56	24.53	14.96
Ba/Rb	14.35	14.61	8.40	12.74	12.35	16.16

INTERMEDIATE AMPHIBOLITES (DIORITE−TONALITE)

Sample number	58250	58258	58251	58262	M3	M10
Locality	Lower County Quarry	Westminster Quarry	Gardiners Quarry	Swinyard Hill summit*	Tank Quarry N Malvern	Tank Quarry N Malvern
Lithology	Diorite	Diorite	Diorite	Tonalite	Tonalite	Tonalite sheet
Grid reference	SO 7678 4468	SO 7656 4608	SO 7664 4209	SO 7618 3861	SO 7682 4706	SO 7682 4706
SiO_2	52.78	47.11	52.60	65.15	60.64	65.26
TiO_2	1.75	1.64	1.67	.66	.68	.45
Al_2O_3	16.09	16.48	16.32	15.43	17.13	17.13
Fe_2O_3	2.84	4.18	3.99	1.65	6.37	3.56
FeO	5.07	7.45	7.00	2.65	—	—
MnO	.14	.23	.20	.08	.13	.08
MgO	3.87	6.19	4.90	1.86	2.93	1.51
CaO	7.11	5.74	3.24	4.31	3.51	3.39
Na_2O	3.91	2.83	3.61	3.90	3.34	4.35
K_2O	1.82	3.24	2.27	2.01	2.70	2.27
P_2O_5	1.07	.94	.17	.19	.29	.18
LOI	2.63	3.37	3.11	1.85	2.10	1.43
Rest			.23	.22	.18	.25
Total	99.40	99.80	99.31	99.96	100.00	99.86
Trace elements in parts per million						
Ba	412	1438	525	923	271	737
Ce	119	147	70	79	44	34
Co	22	42	39	13	19	6
Cr	40	88	151	16	140	79
Cu	21	58	39	12	7	12
Nb	30	24	14	9	6	3
Ni	26	40	46	5	8	3
Pb	8	11	9	12	11	9
Rb	59	103	72	56	82	75
Sr	1407	499	213	394	444	788
Th	1	8	7	6	6	1
V	128	301	198	67	95	52
Y	24	66	40	15	15	7
Zn	98	168	130	45	76	64
Zr	194	261	233	197	193	178
Mg′	47.49	49.59	45.19	44.49	47.67	45.65
Nb/Y	1.25	.36	.35	.60	.40	.43
Zr/Y	8.08	3.95	5.82	13.13	12.87	25.43
Zr/Nb	6.47	10.88	16.64	21.89	32.17	59.33
Zr/TiO_2*10 000	.01	.02	.01	.03	.03	.04
Ce/Y	4.96	2.23	1.75	5.27	2.93	4.86
(Ce/Y)n	12.40	5.57	4.38	13.17	7.33	12.14
Th/Nb	.03	.33	.50	.67	1.00	.33
K/Rb	256.06	261.12	261.71	297.95	273.33	251.24
K/Ba	36.67	18.70	35.89	18.08	82.70	25.57
Ba/Rb	6.98	13.96	7.29	16.48	3.30	9.83

* Tewkesbury (216) Sheet

METAMORPHOSED GRANITES

Sample number	58260	58264	58257	58265	58267	M7
Locality	N Malvern Quarries	N Malvern Quarries	Upper County Quarry	Gullet* Quarry	Dingle Quarry	Tolgate Quarry
Lithology	Granite	Granite	Syeno-granite	Syeno-granite	Granite	Granite
Grid reference	SO 7715 4685	SO 7690 4700	SO 7680 4480	SO 7625 3808	SO 7654 4567	SO 7694 4395
SiO_2	66.29	69.82	70.69	75.33	76.39	74.23
TiO_2	.45	.36	.46	.69	.13	.16
Al_2O_3	17.14	14.78	14.77	10.85	13.15	13.79
Fe_2O_3	1.44	1.41	1.31	1.47	.49	1.54
FeO	2.04	2.28	1.71	2.19	.76	—
MnO	.04	.04	.05	.04	.01	.07
MgO	1.43	1.29	.96	1.23	.74	.47
CaO	2.93	2.35	1.38	.57	1.79	1.04
Na_2O	3.93	3.44	4.12	2.92	3.76	4.08
K_2O	2.99	2.14	3.32	3.05	1.79	3.81
P_2O_5	.19	.10	.12	.05	.03	.04
LOI	1.16	1.34	1.24	1.52	.95	.65
Rest	.28	.30	.17	.21	.18	.15
Total	100.31	99.65	100.30	100.12	100.17	100.03
Trace elements in parts per million						
Ba	1020	1367	756	687	752	691
Ce	39	46	31	79	34	20
Co	14	15	9	13	11	3
Cr	9	10	19	76	16	69
Cu	11	128	6	9	25	3
Nb	7	5	7	13	6	7
Ni	3	7	6	18	5	—
Pb	9	13	8	61	26	19
Rb	76	61	84	88	62	108
Sr	877	618	239	119	482	191
Th	—	3	7	13	10	11
V	54	57	32	52	22	10
Y	7	10	9	31	8	21
Zn	50	41	34	25	15	25
Zr	200	153	170	394	61	82
Mg′	43.31	39.31	37.19	38.42	52.34	36.67
Nb/Y	1.00	.50	.78	.42	.75	.33
Zr/Y	28.57	15.30	18.89	12.71	7.63	3.90
Zr/Nb	28.57	30.60	24.29	30.31	10.17	11.71
Zr/TiO_2*10 000	.04	.04	.04	.06	.05	.05
Ce/Y	5.57	4.60	3.44	2.55	4.25	.95
(Ce/Y)n	13.93	11.50	8.61	6.37	10.63	2.38
Th/Nb		.60	1.00	1.00	1.67	1.57
K/Rb	326.58	291.22	328.09	287.71	239.66	292.84
K/Ba	24.33	12.99	36.45	36.85	19.76	45.77
Ba/Rb	13.42	22.41	9.00	7.81	12.13	6.40

* Tewkesbury (216) Sheet

PEGMATITES			FELSIC MYLONITES (GRANITIC PROTOLITH)	
Sample number	M4	M8	58255	58268
Locality	Gullet* Quarry	Tank Quarry	Gullet* Quarry	Hollybush* Quarry
Lithology	Pegmatite vein	Pegmatite vein	Felsic mylonite	Felsic mylonite
Grid reference	SO 7618 3807	SO 7682 4706	SO 7625 3808	SO 7593 3718
SiO_2	75.20	74.25	69.44	70.03
TiO_2	.02	.02	.78	.08
Al_2O_3	12.48	13.92	14.38	13.16
Fe_2O_3	.37	.74	1.73	.44
FeO	—	—	2.59	.59
MnO	.02	.02	.05	.05
MgO	.06	.35	1.55	.48
CaO	1.72	1.01	.52	4.66
Na_2O	2.98	3.64	2.96	5.08
K_2O	5.40	5.39	3.57	2.11
P_2O_5	.02	.01	.08	.03
LOI	1.46	.45	1.79	3.03
Rest	.10	.08	.22	.11
Total	99.83	99.91	99.66	99.85
Trace elements in parts per million				
Ba	490	179	810	448
Ce	—	—	55	—
Co	—	—	10	6
Cr	77	80	86	21
Cu	4	8	17	5
Nb	—	8	16	11
Ni	3	5	24	8
Pb	28	27	45	12
Rb	122	115	97	59
Sr	96	158	135	230
Th	1	5	11	2
V	—	8	66	22
Y	8	19	28	15
Zn	—	—	44	5
Zr	36	51	331	73
Mg'	24.31	48.37	39.98	46.45
Nb/Y		.42	.57	.73
Zr/Y	4.50	2.68	11.82	4.87
Zr/Nb		6.38	20.69	6.64
Zr/TiO_2*10 000	.18	.10	.04	.09
Ce/Y			1.96	
(Ce/Y)n			4.91	
Th/Nb		.63	.69	.18
K/Rb	367.42	389.06	305.51	296.87
K/Ba	91.48	249.96	36.59	39.10
Ba/Rb	4.02	1.56	8.35	7.59

* Tewkesbury (216) Sheet

2. Major and trace element data for late Precambrian microdiorite intrusions (Dearnley, 1990)

MICRODIORITE DYKES

Sample number	58239	58240	58243	58254
Locality	N Malvern Quarries	N Malvern Quarries	Upper County Quarry	Gullet* Quarry
Lithology	Microdiorite	Microdiorite	Microdiorite	Microdiorite
Grid reference	SO 7618 3807	SO 7682 4706	SO 7673 4477	SO 7625 3808
SiO_2	48.81	49.94	47.84	51.26
TiO_2	3.28	2.61	2.85	2.68
Al_2O_3	13.37	13.67	13.52	14.19
Fe_2O_3	4.50	4.21	4.39	4.25
FeO	8.98	8.33	9.33	8.25
MnO	.22	.24	.24	.16
MgO	4.64	4.74	5.56	4.74
CaO	6.04	6.07	8.36	4.39
Na_2O	4.01	3.83	3.79	4.09
K_2O	1.03	1.20	.79	1.03
P_2O_5	.65	.33	.39	.36
LOI	2.05	2.14	1.90	3.46
Rest	.26	.23	.22	.18
Total	97.84	97.54	99.18	99.04
Trace elements in parts per million				
Ba	481	519	308	257
Ce	84	44	35	40
Co	29	37	49	30
Cr	52	46	78	52
Cu	19	22	55	25
Nb	8	6	8	8
Ni	26	22	36	16
Pb	12	13	5	10
Rb	30	39	26	21
Sr	218	272	253	162
Th	4	3	—	5
V	234	295	355	297
Y	86	57	56	47
Zn	151	129	109	118
Zr	570	231	240	225
Mg'	38.82	41.07	42.73	41.16
Nb/Y	.09	.11	.14	.17
Zr/Y	6.63	4.05	4.29	4.79
Zr/Nb	71.25	38.50	30.00	28.13
Zr/TiO_2*10 000	.02	.01	.01	.01
Ce/Y	.98	.77	.63	.85
(Ce/Y)n	2.44	1.93	1.56	2.13
Th/Nb	.50	.50		.63
K/Rb	285	255.42	252.22	407.14
K/Ba	17.78	19.19	21.29	33.27
Ba/Rb	16.03	13.31	11.85	12.24

* Tewkesbury (216) Sheet

3. Major and trace element data for samples from the Kempsey Borehole (Pharaoh and Brewer, unpublished data)

KEMPSEY BOREHOLE [SO86094933]

Sample number	KEM1	KEM2	KEM3	KEM4	KEM5	KEM6	KEM7
Lithology	Intermediate clast	Intermediate matrix	Intermediate clast	Intermediate clast	Intermediate clast	Lithic sandstone	Lithic sandstone
Depth (m)	2338.3	2338.3	2338.3	2337.5	2336.6	2461.9	3009.6
SiO_2	55.85	55.08	56.14	56.25	54.47	52.93	48.13
TiO_2	.81	.78	.62	.78	.76	1.00	.93
Al_2O_3	17.04	16.64	17.63	16.60	16.72	18.66	21.22
Fe_2O_3	8.83	8.82	7.71	8.58	9.12	10.00	9.38
MnO	.16	.16	.15	.17	.16	.18	.13
MgO	4.56	4.80	4.14	4.70	5.22	6.06	7.56
CaO	.77	1.05	1.11	.93	1.20	1.56	.61
Na_2O	2.54	2.85	1.96	2.80	2.01	2.37	4.64
K_2O	5.72	5.80	7.16	5.39	5.85	2.86	2.50
P_2O_5	.31	.32	.24	.30	.32	.28	.22
LOI	3.34	3.74	3.10	3.58	4.57	4.50	5.08
Rest	.32	.30	.42	.34	.28	.20	.14
Total	100.25	100.34	100.38	100.42	100.68	100.60	100.54
Trace elements in parts per million							
Ba	1870	1782	2825	1929	1517	881	503
Ce	5	5	8	5	7	20	26
Co	24	20	24	23	21	30	21
Cr	77	30	58	147	58	82	31
Cu	59	63	52	70	73	44	16
Nb	1	3	1	2	1	4	7
Ni	3	3	4	4	6	13	8
Pb	6	3	4	2	8	7	3
Rb	193	194	257	176	204	73	74
Sr	98	98	96	118	85	61	89
Th	5	3	4	5	3	5	12
V	157	160	132	160	172	152	95
Y	21	17	19	24	24	28	26
Zn	92	91	88	96	97	112	111
Zr	61	63	51	61	59	106	122
Mg'	50.66	51.87	51.54	52.03	53.13	54.55	61.48
Nb/Y	.05	.18	.05	.08	.04	.14	.27
Zr/Y	2.90	3.71	2.68	2.54	2.46	3.79	4.69
Zr/Nb	61.00	21.00	51.00	30.50	59.00	26.50	17.43
Zr/TiO_2*10 000	.01	.01	.01	.01	.01	.01	.01
Ce/Y	.24	.29	.42	.21	.29	.71	1.00
(Ce/Y)n	.60	.74	1.05	.52	.73	1.79	2.50
Th/Nb	4.55	1.00	4.00	2.50	3.00	1.25	1.71
K/Rb	246.02	248.17	231.27	254.22	238.04	325.22	280.44
K/Ba	25.39	27.02	21.04	23.19	32.01	26.95	41.26
Ba/Rb	9.69	9.19	10.99	10.96	7.44	12.07	6.80

APPENDIX 3

Boreholes

Selected boreholes in Worcester district

This list gives the permanent BGS record number, name, National Grid reference, date drilled, surface level (SL; where known), stratigraphical summary and total depth of selected boreholes referred to in this memoir and which lie within the 1:50 000 Worcester Sheet. Other published references are also given. Records of these and other non-confidential boreholes can be consulted at BGS, Keyworth and copies obtained at a fixed tariff.

SO74NE/15   Malvern Link No. 1 [7887 4922]. 1903. Mercia Mudstone Group to 214.94m, Bromsgrove Sandstone Formation drilled to 267.61 m. Richardson (1930).

SO74NE/16   Malvern Link No. 2 [7882 4917]. 1905. Mercia Mudstone Group to 214.88m, Bromsgrove Sandstone Formation drilled to 289.56 m. Richardson (1930).

SO74SE/2   Colwall Borehole [7572 4265]. 1890–94. SL c.+137.16 m. Drift (Head) to 6.10 m, Raglan Mudstone Formation to 353.56, Downton Castle Sandstone to 358.44 m, Upper Ludlow Shale drilled to 379.17 m. Richardson (1935).

SO84NE/2   Kempsey (No. 1) Borehole [8609 4933]. 1979. SL +19.82 m. Drift to c.7.62 m, Mercia Mudstone Group to 348.47 m, Bromsgrove Sandstone Formation to 421.65 m, Wildmoor Sandstone Formation to 915.55 m, Kidderminster Formation to 1200.3 m, Bridgnorth Sandstone Formation to 1367.07 m, Kempsey Formation drilled to 3012.19 m. Whittaker (1980).

SO84SW/1   Upton upon Severn Borehole [8341 4043]. 1906–12. SL +48.78 m. Mercia Mudstone Group to 405.08 m, Bromsgrove Sandstone Formation drilled to 518.16 m. Richardson (1923, 1930).

SO85NE/23   Worcester (Heat Flow) Borehole [8624 5762]. 1983. SL +c.30. Mercia Mudstone Group to 295.47 m, Bromsgrove Sandstone Formation drilled to 301.32 m.

SO94SW/1   Defford Borehole [9142 4299]. 1934. Drift to ?4.88 m, Lias Group to ?170.88 m, Penarth Group to ?179.22 m, Mercia Mudstone Group drilled to 266.09 m.

Boreholes in adjacent areas

The following lists the permanent BGS record number, name, National Grid reference and 1:50 000 geological sheet number of boreholes referred to in this memoir which lie outside the area of the 1:50 000 Worcester Sheet.

SO66SW/1   Collington Borehole [SO 6460 6100]. Sheet 182.

SO73SE/6   Eldersfield Borehole [SO 7891 3221]. Sheet 216.

SO82SE/49   Staverton Borehole [SO 8840 2290]. Sheet 216.

SO83NE/5   Twyning Borehole [SO 8943 3664]. Sheet 216.

SO86SE/13   Westwood Borehole [SO 8654 6343]. Sheet 182

SO87NE/3   Bellington No. 2 Borehole [SO 8769 7679]. Sheet 182.

SO93NE/1   Bredon Hill No. 1 (Lalu Barn) Borehole [SO 9577 3996]. Sheet 200.

SO93NE/2   Bredon Hill No. 2 (Scarborough Cottages) [SO 9643 3788]. Sheet 217.

SO94SE/1   Netherton Borehole [SO 9982 4075]. Sheet 200.

SO96NE/30   Sugarbrook No. 3 Borehole [SO 9593 6810]. Sheet 183.

SO96SW/1   Saleway Borehole [9285 6017]. Sheet 182.

SO97SW/5   Whitford Mill [SO 944 707]. Sheet 182.

SP01SE/1   Stowell Park Borehole [SP 084 118]. Sheet 235.

SP06NW/100   Webheath No.1 Borehole [SP 0098 6693]. Sheet 183.

SP06NE/1   Putcheons Farm Borehole [SP 0694 6577]. Sheet 183.

SP17NE/184   Knowle Borehole [SP 1890 7773]. Sheet 183.

SP25SW/1   Knights Lane [SP 2242 5497]. Sheet 200.

SP47SW/72   Home Farm Borehole [SP 4317 7309]. Sheet 184.

SU09SW/52   Cooles Farm Borehole [SU 0164 9213]. Sheet 252.

APPENDIX 4

Geological Survey photographs

Eighty seven photographs illustrating the geology of the Worcester district are deposited for reference in the headquarters of the British Geological Survey, Keyworth, Nottingham NG12 5GG; in the library at the BGS, Murchison House, West Mains Road, Edinburgh EH9 3LA; and in the BGS Information Office at the Natural History Museum Earth Galleries, Exhibition Road, London SW7 2DE. They belong to the A Series and are numbered as follows, with the years photographed:

A3242–7	1925
A6230–42	1933
A8455–66	1950
A11099	1969
A11121–3	1969

A13615	1980
A13939	1982
A13968–72	1982
A15067–94	1991
A15376–87	1991

The photographs depict details of the various rocks and sediments exposed and also include general views and scenery. Some of the older photographs include rock exposures which are no longer visible. A list of titles can be supplied on request. The photographs can be supplied at a fixed tariff, as black and white or colour prints and 2×2 colour transparencies for the 1969 and later ones; those shot in 1925, 1933 and 1950 are available in black and white only.

AUTHOR CITATIONS FOR FOSSIL SPECIES

Chapter 4   Silurian

Ammonidium microcladum (Downie) Lister, 1970

Amphistrophia funiculata (McCoy, 1846)

Atrypa reticularis Linnaeus, 1758)

Auchenaspis egertoni Larkester, 1870

Calymene puellaris Reed, 1920 [formerly *C. neointermedia* Richter and Richter, 1954]

Camarotoechia nucula (J de C Sowerby, 1839)

Cardiola interrupta (J de C Sowerby, 1839)

Coolinia applanata (Salter, 1846)

Coolinia pecten (Linnaeus, 1758)

Coryssognathus dubius (Rhodes, 1953)

Costistricklandia lirata (J de C Sowerby, 1839)

Craniops implicatus (J de C Sowerby, 1839)

Cymatiosphaera octoplana Downie, 1959

Cypricardinia subplanulata Reed, 1928

Cyrtia exporrecta (Wahlenberg, 1818)

Cytherellina siliqua (Jones, 1855)

Dalejina hybrida (J de C Sowerby, 1839)

Dalmanites myops (König, 1825)

Dayia navicula (J de C Sowerby, 1839)

Deunffia brevispinosa Downie, 1960

Dicoelosia biloba (Linnaeus, 1758)

Diexallophasis denticulata (Stockmans & Willière) Loeblich, 1970

Diexallophasis pachymura (Hill) Dorning, 1981

Dilatisphaera dameryensis Dorning, 1981

Distomodus staurognathoides (Walliser, 1964)

Domasia amphora Martin, 1969

Domasia bispinosa Downie, 1960

Domasia limaciformis (Stockmans & Willière) Cramer, 1970

Domasia trispinosa Downie, 1960

Encrinurus diabolus Tripp, Temple and Gass, 1977

Encrinurus stubblefieldi Tripp, 1962

Eocoelia angelini (Lindström, 1861)

Eocoelia curtisi Ziegler, 1966

Eocoelia hemisphaerica (J de C Sowerby, 1839)

Eocoelia sulcata (Prouty, 1923)

Eoplectodonta duvalii (Davidson, 1847)

Favosites gothlandicus Lamarck, 1816

Frostiella groenvalliana Martinsson, 1963

Fuchsella amygdalina (J de C Sowerby, 1839)

Goniophora cymbaeformis (J de C Sowerby, 1839)

Gracilisphaeridium encantador Cramer, 1970

Gypidula lata Alexander, 1936

Hemiaechminoides monospinus Morris and Hill   1952

Howellella elegans (Muir-Wood, 1925)

Howellella subinsignis (Reed, 1927) [as *Delthyris elevata*]

Hyattidina canalis (J de C Sowerby, 1839)

Icriodella inconstans Aldridge, 1972

Icriodella malvernensis Aldridge, 1972

Isorthis orbicularis (J de C Sowerby, 1839)

Jonesea [Aegiria] grayi (Davidson, 1849)

Kockelella ranuliformis (Walliser, 1964)

Leptaena depressa (J de C Sowerby, 1824)

Leptaena rhomboidalis (Wilckens, 1769)

Leptostrophia filosa (J de C Sowerby, 1839)

Lingula lewisii (J de C Sowerby, 1839)

Lingula minima (J de C Sowerby, 1839)

Londinia fissurata Shaw, 1969

Menoeidina lavoiei Copeland   1974

Meristina obtusa (J Sowerby, 1818)

Microsphaeridiorhynchus nucula (J de C Sowerby, 1839)

Neobeyrichia lauensis (Kiesow, 1888)

Nuculites antiquus (J de C Sowerby, 1839)

Orbiculoidea rugata (J de C Sowerby, 1839)

Orthonota rigida (J de C Sowerby, 1839)

Ozarkodina aldridgei Uyeno in Uyeno & Barnes, 1983

Ozarkodina bohemica bohemica (Walliser, 1964)

Ozarkodina confluens (Branson & Mehl, 1933)

Ozarkodina excavata(Branson & Mehl, 1933)

Ozarkodina gulletensis (Aldridge, 1972)

Ozarkodina remscheidensis eosteinhornensis (Walliser, 1964)

Panderodus equicostatus (Rhodes, 1953)

Pentamerus oblongus J de C Sowerby, 1839

Plectatrypa imbricata (J de C Sowerby, 1839)

Poleumita globosa (Schlotheim, 1820)

Pristiograptus tumescens (Wood, 1900)

Protochonetes ludloviensis Muir-Wood, 1962

Protochonetes minimus (J de C Sowerby, 1839)

Pterospathodus amorphognathoides Walliser, 1964

Resserella canalis (J de C Sowerby, 1839) [as *R. elegantula* Dalman]

Resserella sabrinae sabrinae Bassett, 1972

Resserella [Parmothis] elegantula (Dalman, 1827)

Rhynchotreta cuneata (Dalman, 1828)

Saetograptus (Saetograptus) chimaera (Barrande, 1850)

Saetograptus (Colonograptus) varians (Wood, 1900)

Saetograptus (S.) leintwardinensis (Lapworth, 1880)

Salopidium granuliferum (Downie) Dorning, 1981

Salopina lunata (J de C Sowerby, 1839)

Schlotheimophyllum patellatum (Schlotheim, 1820)

Serpulites longissimus J de C Sowerby, 1839

Shagamella minor (Salter, 1848) [as *Chonetes lepisma* (J de C Sowerby, 1839)]

Shaleria ornatella (Davidson, 1871)

Sphaerirhynchia wilsoni (J Sowerby, 1816)

Spongarium aequistriatum McCoy, 1850

Stegerhynchus borealis (von Buch, 1834)

Stricklandia lens (J de C Sowerby, 1839)

Strophonella euglypha (Dalman, 1828)

Tylotopalla digitifera Loeblich, 1970

Tylotopalla robustispinosa (Downie) Eisenack, Cramer & Díez, 1973

Chapter 5   Devonian

Beaconites antarcticus Vialov, 1962

Benneviaspis anglica Stensiö, 1932

Benneviaspis lankesteri Stensiö, 1932

Cephalaspis agassizi (Lankester, 1870)

Cephalaspis cradleyensis Stensiö, 1932

Cephalaspis fletti Stensiö, 1932

Cephalaspis langi Stensiö, 1932

Cephalaspis salweyi Egerton, 1857

Cephalaspis sollasi Stensiö, 1932

Carditomantea reecei Reed, 1934

Carditomantea virelyi (Barrois, Pruvost and Dubois, 1920

Cephalaspis lyelli Agassiz, 1835

Didymaspis grindrodi Lankester, 1867

Eurymyella ammonis Reed, 1934

Eurymyella shaleri var. *longa* Williams, 1912

Modiolopsis gradata (Salter, 1848)

Modiolopsis leightoni var. *curta* Reed, 1934

Modiolopsis nilssoni (Hisinger, 1837)

Modiolopsis complanata trimpleyensis Reed, 1934

Pteraspis crouchii Lankester, 1864

Pteraspis leathensis King, 1934, nom. nud

Pteraspis rostratus (Agassiz, 1835)

Pterygotus anglicus Agassiz, 1844

Scaphaspis lloydii (Agassiz, 1835)

Chapter 7   Triassic

Abietineaepollenites dunrobinensis Couper, 1958

Acanthotriletes ovalis Nilsson, 1958

Acanthotriletes varius Nilsson, 1958

Alisporites circulicorpus Clarke, 1965

Alisporites grauvogeli Klaus, 1964

Alisporites microreticulatus Reinhardt, 1964

Alisporites parvus de Jersey, 1964

Alisporites thomasii (Couper) Nilsson, 1958

Alisporites toralis (Leschilc) Clarke, 1965

Angustisulcites gorpii Visscher, 1966

Angustisulcites grandis (Freudenthal) Visscher, 1966

Angustisulcites klausii Freudenthal, 1964

Apiculatasporites plicatus Visscher, 1966

Aratrisporites fimbriatus (Klaus) Mädler emend. Morbey, 1975

Aratrisporites granulatus (Klaus) Playford & Dettmann, 1965

Aratrisporites saturni (Thiergart) Mädler, 1964

Avicula contorta Portlock, 1843

Beaumontella caminuspina (Wall) Below, 1987

Beaumontella langii (Wall) Below, 1987

Brodispora striata Clarke, 1965

Calamospora mesozoica Couper, 1958

Camarozonosporites rudis (Leschik) Klaus, 1960

Camerosporites secatus Leschik emend. Scheuring, 1978

Carnisporites anteriscus Morbey, 1975

Carnisporites lecythus Morbey, 1975

Carnisporites leviornatus (Levet-Carette) Morbey, 1975

Carnisporites spiniger (Leschik) Morbey, 1975

Chasmatosporites magnolioides (Erdtman) Nilsson, 1958

Chlamys valoniensis (Defrance, 1825)

Cingulizonates rhaeticus (Reinhardt) Schulz, 1967

Classopollis torosus (Reissinger) Balme, 1957

Colpectopollis ellipsoideus Visscher, 1966

Contignisporites problematicus (Couper) Döring, 1965

Converrucosisporites cameroni (de Jersey) Playford & Dettmann, 1965

Converrucosisporites luebbenensis Schulz, 1967

Convolutispora microrugulata Schulz, 1967

Cornutisporites seebergensis Schulz, 1962

Cuneatisporites radialis Leschik emend. Scheuring, 1978

Cyathiditis australis Couper, 1953

Cyclotriletes microgranifer Mädler, 1964

Cyclotriletes oligogranifer Mädler, 1964

Cymatiosphiaera polypartita Morbey, 1975

Dacryomya titei (Moore, 1861)

Dapcodinium priscum Evitt, 1961

Deltoidospora neddeni (Potonié) Orbell, 1973

Densoisporites nejburgii (Schulz) Balme, 1970

Duplicisporites granulatus Leschik emend. Scheuring, 1970

Duplicisporites scurrilis (Scheuring) Scheuring, 1978

Duplicisporites verrucosus Leschik emend. Scheuring, 1978

Ellipsovelatisporites plicatus Klaus, 1960

Ellipsovelatisporites rugosus Scheuring, 1970

Enzonalasporites vigens Leschik, 1955

Eotrapezium concentricum (Moore, 1861)

Eotrapezium germari (Dunker, 1846)

Euestheria minuta (Zieten, 1833)

Geopollis zwolinskai (Lund) Brenner, 1986

'Gervillia' praecursor Quenstedt, 1858

Gliscopollis meyeriana (Klaus) Venkatachala, 1966

Granuloperculatipollis rudis Venkatachala & Góczán emend. Morbey, 1975

Haberkornia gudati Scheuring, 1978

Haberkornia parva Scheuring, 1978

Illinites chitonoides Klaus, 1964

Illinites kosankei Klaus, 1964

Klausipollenites devolvens (Leschik) Clarke, 1965

Kraeuselisporites hoofddijkensis Visscher, 1966

Kraeuselisporites reissingeri (Harris) Morbey, 1975

Kyrtomisporis laevigatus Mädler, 1964

Labiisporites granulatus Leschik, 1956

Leptolepidites argenteaeformis (Bolkhovitina) Morbey, 1975

Limbosporites lundbladii Nilsson, 1958

Lunatisporites acutus Leschik emend. Scheuring, 1970

Lunatisporites pellucidus (Goubin) Helby ex. de Jersey, 1972

Lunatisporites rhaeticus (Schulz) Warrington, 1974

Lycopodiacidites rhaeticus Schulz, 1967

Lyriomyophoria postera (Quenstedt, 1856)

Masculostrobus bromsgrovensis Grauvogel-Stamm, 1972

Masculostrobus willsi Townrow, 1962

Michrystridium licroidium Morbey, 1975

Michrystridium lymense var. *gliscum* Wall, 1965

Michrystridium lymense var. *lymense* Wall, 1965

Michrystridium lymense var. *rigidum* Wall, 1965

Microreticulatisporites fuscus (Nilsson) Morbey, 1975

'Modiolus' sodburiensis Vaughan, 1904

'Natica' oppelii Moore, 1861

Nevesisporites bigranulatus (Levet-Carette) Morbey, 1975

Osmundacidites alpinus Klaus, 1960

Osmundacidites wellmanii Couper, 1953

Ovalipollis pseudoalatus (Thiergart) Schuurman, 1976

Partitisporites quadruplicis (Scheuring) Van der Eem, 1983

Parvisaccites triassicus Scheuring, 1978

Patinasporites densus Leschik emend. Scheuring, 1970

Perinopollenites elatoides Couper, 1958

Perinosporites thuringiacus Schulz, 1962

Perotrilites minor (Mädler) Antonescu & Taugourdeau Lantz, 1973

Plaesiodictyon mosellanum Wille, 1970

Podosporites amicus Scheuring, 1970

Polycingulatisporites bicollateralis (Rogalska) Morbey, 1975

Porcellispora longdonensis (Clarke) Scheuring emend. Morbey, 1975

Praecirculina granifer (Leschik) Klaus, 1960

Protocardia rhaetica (Merian, 1853)

Protodiploxypinus doubingeri (Klaus) Warrington, 1974

Protodiploxypinus fastidiosus (Jansonius) Warrington, 1974

Protodiploxypinus potoniei (Mädler) Scheuring, 1970

Protodiploxypinus sittleri (Klaus) Scheuring, 1970

Protohaploxypinus microcorpus (Schaarschmidt) Clarke, 1965

Quadraeculina anellaeformis Maljavkina, 1949

Retisulcites perforatus (Mädler) Scheuring, 1970

Retitriletes austroclavatidites (Cookson) Döring, Krutzsch, Mai & Schulz, 1963

Rhaetavicula contorta (Portlock, 1843)

Rhaetipollis germanicus Schulz, 1967

Rhaetogonyaulax rhaetica (Sarjeant) Loeblich and Loeblich emend. Below, 1987

Ricciisporites tuberculatus Lundblad, 1954

Ricciisporites umbonatus Felix & Burbridge, 1977

Rimaesporites potoniei Leschik, 1955

Staurosaccites quadrifidus Dolby *in* Dolby & Balme, 1976

Stellapollenites thiergartii (Mädler) Clement-Westerhof, Van der Eem, Van Erve, Klasen, Schuurman & Visscher, 1974

Stereisporites radiatus Schulz, 1970

Striatoabieites balmei Klaus emend. Scheuring, 1978

Sulcatisporites kraeuseli Mädler, 1964

Todisporites major Couper, 1958

Todisporites minor Couper, 1958

Triadispora aurea Scheuring emend. Scheuring, 1978

Triadispora barbata Scheuring, 1978

Triadispora crassa Klaus, 1964

Triadispora epigona Klaus, 1964

Triadispora falcata Klaus, 1964

Triadispora obscura Scheuring, 1970

Triadispora plicata Klaus, 1964

Triadispora stabilis Scheuring emend. Scheuring, 1978

Triadispora vilis Scheuring, 1970

Triancoraesporites ancorae (Reinhardt) Schulz, 1967

Triancoraesporites reticulatus Schulz, 1962

Tsugaepollenites oriens Klaus, 1964

Tsugaepollenites pseudomassulae (Mädler) Morbey, 1975

Tutcheria cloacina (Quenstedt, 1856)

Vallasporites ignacii Leschik, 1955

Verrucosisporites contactus Clarke, 1965

Verrucosisporites jenensis Reinhardt & Schmitz ex. Reinhardt, 1964

Verrucosisporites krempii Mädler, 1964

Verrucosisporites morulae Klaus, 1960

Verrucosisporites remyanus Mädler, 1964

Vesicaspora fuscus (Pautsch) Morbey, 1975

Voltziaceaesporites heteromorpha Klaus 1964

Zebrasporites interscriptus (Thiergart) Klaus, 1960

Zebrasporites laevigatus (Schulz) Schulz, 1967

Chapter 8   Jurassic

Acanthotriletes varius Nilsson, 1958
Ammonites jamesoni J de C Sowerby, 1827
Ammonites planorbis J de C Sowerby, 1824
Asteroceras suevicum (Quenstedt, 1884)
Calamospora mesozoica Couper, 1958
Cardinia ovalis (Stutchbury, 1842)
Carnisporites anteriscus Morbey, 1975
Carnisporites spiniger (Leschik) Morbey, 1975
Classopollis torosus (Reissinger) Balme, 1957
Cleviceras elegans (J Sowerby, 1815)
Cleviceras exaratum (Young & Bird, 1828)
Contignisporites problematicus (Couper) Döring, 1965
Epophioceras longicella (Quenstedt, 1883)
Gryphaea arcuata Lamarck, 1801
Gryphaea cymbium Lamarck, 1819
Hippopodium ponderosum J Sowerby, 1819
Kraeuselisporites reissingeri (Harris) Morbey, 1975
Liostrea hisingeri (Nilsson, 1831)
Micrhystridium lymense var. *gliscum* Wall, 1965

Micrhystridium lymense var. *lymense* Wall, 1965
Osmundacidites wellmanii Couper, 1953
Oxynoticeras oxynotum (Quenstedt, 1843)
Plagiostoma giganteum J Sowerby, 1814
Pleuroceras spinatum (Bruguière, 1789)
Pleuromya costata (Young & Bird, 1828)
Porcellispora longdonensis (Clarke) Scheuring emend. Morbey, 1975
Pseudopecten equivalvis (J de C Sowerby, 1816)
Psiloceras planorbis (J de C Sowerby, 1824)
Quadraeculina anellaeformis Maljavkina, 1949
Schlotheimia angulata (Schlotheim, 1820)
Slatterites slatteri (Wright, 1982)
Tasmanites newtoni Wall, 1965
Tasmanites suevicus (Eisenack) Wall, 1965
Tetrarhynchia dumbletonensis (Davidson, 1878)

Chapter 9   Quaternary

Ancylus fluviatilis (Müller, 1774)
Bithynia tentaculata (Linnaeus, 1758)
Cervus elaphas Linnaeus, 1758

Coelodonta antiquitatis (Blumenbach, 1799)
Corbicula fluminalis (Müller, 1774)
Discus rotundatus (Müller, 1774)
Equus spelaeus Owen, 1869
Groenlandia densa (Linneaus, 1753) Fourreau, 1869
Gryphaea arcuata Lamarck, 1801 [derived Jurassic]
Hippopotamus amphibius Linnaeus, 1758
Hippopotamus major Cuvier, 1824
Limnocythere sanctipatricii Brady & Robertson, 1869
Mammuthus primigenius (Blumenbach, 1799)
Palaeoloxodon antiquus (Falconer & Cautley, 1847)
Pisidium amnicum (*Cyclas amnica*) (Müller, 1774)
Pisidium casertanum Poli, 1791)
Pisidium henslowanum (Sheppard, 1823)
Pisidium moitessierianum (Paladilhe, 1866)
Pisidium vincentianum Woodward, 1913
Potamogeton praelongus Wulfen, 1805
Rangifer tarandus (Linnaeus, 1758)
Sphaerium corneum (Linnaeus, 1758
Trichia hispida (Linnaeus, 1758
Valvata piscinalis (Müller, 1774)

INDEX

BRITISH GEOLOGICAL SURVEY

Keyworth, Nottingham NG12 5GG
0115 936 3100

Murchison House, West Mains Road, Edinburgh
EH9 3LA   0131-667 1000

London Information Office, Natural History Museum
Earth Galleries, Exhibition Road, London SW7 2DE
0171-589 4090

The full range of Survey publications is available through the Sales Desks at Keyworth and at Murchison House, Edinburgh, and in the BGS London Information Office in the Natural History Museum (Earth Galleries). The adjacent bookshop stocks the more popular books for sale over the counter. Most BGS books and reports can be bought from The Stationery Office and through The Stationery Office agents and retailers. Maps are listed in the BGS Map Catalogue, and can be bought together with books and reports through BGS-approved stockists and agents as well as direct from BGS.

The British Geological Survey carries out the geological survey of Great Britain and Northern Ireland (the latter as an agency service for the government of Northern Ireland), and of the surrounding continental shelf, as well as its basic research projects. It also undertakes programmes of British technical aid in geology in developing countries as arranged by the Overseas Development Administration.

The British Geological Survey is a component body of the Natural Environment Research Council.

The Stationery Office publications are available from:

The Stationery Office Publications Centre
(Mail, fax and telephone orders only)
PO Box 276, London SW8 5DT
Telephone orders 0171-873 9090
General enquiries 0171-873 0011
Queuing system in operation for both numbers
Fax orders 0171-873 8200

The Stationery Office Bookshops
49 High Holborn, London WC1V 6HB
(counter service only)
0171-873 0011   Fax 0171-831 1326
68–69 Bull Street, Birmingham B4 6AD
0121-236 9696   Fax 0121-236 9699
33 Wine Street, Bristol BS1 2BQ
0117-9264306   Fax 0117-9294515
9 Princess Street, Manchester M60 8AS
0161-834 7201   Fax 0161-833 0634
16 Arthur Street, Belfast BT1 4GD
01232-238451   Fax 01232-235401
71 Lothian Road, Edinburgh EH3 9AZ
0131-228 4181   Fax 0131-229 2734
The Stationery Office Oriel Bookshop,
The Friary, Cardiff CF1 4AA
01222-395548   Fax 01222-384347

The Stationery Office's Accredited Agents
(see Yellow Pages)

And through good booksellers